Beyond the Boundaries of Science helps the readers to think and reflect on the profound question of the origin of the universe and life that has plagued humanity throughout the ages. In this book, Dr. Latha Christie presents some of the enduring and exciting mysteries of science and explores the possibility of a master theory or a "theory of everything," that remains as one of the greatest challenges in physics. She pieces together the current ideas and the evidence while investigating some of the interesting questions like: Did time itself have a beginning, or was our beginning somewhere in the middle of an infinite quantity of time? How to explain the rise of complex, intelligent life? Is this universe imbued with purpose? It is a pleasure to endorse this book as it is sophisticated, simple and arguably groundbreaking.

Prof. N. Balakrishnan, Honorary Professor, Supercomputer Education and Research Centre, Indian Institute of Science, Bangalore, India.

Beyond the Boundaries of Science is a fabulous book, filled with nuggets from deeply digging into the hidden treasures of scientific discoveries, behind the origin of the universe, and the origin of life. In this, Dr. Christie shows how the latest scientific discoveries in various fields, namely, cosmology, physics, palaeontology, archaeology, and biology support the idea that there is a supernatural intelligence behind the fine-tuning, the beginning condition of the universe and life. She also suggests that there is a common ground between science and the creation narrative of Genesis 1. In this wonderful overview of our origins, her ideas are well worth reading, and I highly recommend this work as a great resource.

Dr. S. Christopher, Distinguished Scientist and former Chairman, DRDO, and Secretary, Department of Defense R&D, India; Professor, IIT, Madras.

Dr. Christie has excellent scientific credentials, and I'm glad she's bringing them to the integration of religion and science. The book is compelling, and I wish her all the best in this vital work.

Dr. Greg Cootsona, Professor, California State University, Chico, US, and author of *Mere Science and Christian Faith*

This is a fine attempt to wrestle with an important topic and represents one option among others espoused by evangelical Christians, and Dr. Christie has argued strenuously and carefully for her position.

Dr. Jacob Cherian, Professor of NT and Dean of Faculty, Centre for Global Leadership Development, Bangalore, India.

Fasten your seatbelts. You are in for the ride of your life. In fact, it is the "ride of life." Dr. Latha Christie takes you on a scientific journey from the beginning of time, giving a beautiful and historic summary of the key ideas of science and astronomy demonstrating the limits of our knowledge, but yet the value of true scientific discoveries. She then addresses the origin of humans, succinctly summarizing the major views, both secular and religious. This is a fun book. Even if you are not a scientist, you will grasp the enormity of the issues and the need for honest evaluation of scientific worldviews. All this is to put the Bible and the Lord Jesus into the center of God's plan for the ages. She leaves many questions open, not demanding total agreement, but yet lets her personal journey of faith shine through in the context of scientific investigation. This is an extremely important book for anyone seeking 'truth', but being sidetracked by scientific questions that seem counter to biblical teaching.

Dr. Jerry E. White, Astrodynamicist, Co-author of *Fundamentals of Astrodynamics*, **Major General, US Air Force (retd.), International President Emeritus, The Navigators.**

I consider it a privilege and honor to heartily recommend—*Beyond the Boundaries* to readers of all persuasions. I have known Dr. Christie for some time and, from time to time, we had discussed various aspects of what you now see in print. Her approach to this subject reminds me of the famous dictum of the late American philosopher-apologist, Dr. Francis Schaeffer: "All truth is God's truth!" The book will surely impress both agnostic and Christian readers of the correspondence that exists between faith and evidence. With her academic and scientific background, Dr. Christie weaves her narrative expertly through the maze of documentation to establish the reality that the Christian Scriptures and Science do NOT belong to non-overlapping magisteria (NOMA) but interact with each other in tandem. This book is an exhilarating addition to the study that must occupy the minds of scientists and lay-readers alike.

L.T. Jeyachandran, Chief Engineer (Civil) (retd.), Government of India.

Beyond the Boundaries of Science is a remarkable journey through the most recent scientific findings, providing compelling evidence for a supernatural Creator behind the beginning of our universe and the beginning of life. Dr. Christie's scientific background and theological insight have helped her to explain the evidence in a way that excites awe and wonder. Her model for approaching the creation sequence in Genesis 1 suggests that both Scripture and Science tell the same story. It is my pleasure to endorse this brilliant and insightful book

> Dr. Josh McDowell, award-winning author, apologist,
> founder, Josh McDowell Ministry.

Science asks "WHY?" Technology asks "HOW?" Engineers deal with "Real-Life problems" and therefore have to ask BOTH. If you want to go beyond—to the "Conceptual World," and ask more questions about "WHAT?" (e.g., "What is the origin?", "What is the reason?", "What is the meaning?", etc.)—this book is a good one for you. Dr. Latha Christie asks many more and tries to answer them all.

> Dr. Levent Sevgi, Turkish engineering educator, researcher, and author;
> Professor, Istanbul Okan University, Turkey

In this thought-provoking book, Dr. Christie speaks to those who wonder about the relationship between science and Christianity. Through cutting-edge scientific evidence, she presents the case that, under scrutiny, Scripture and science recite the same story with integrity that can challenge and convince. If you are wrestling with questions about the relationship of modern science with Christian faith and willing to dig deeper into this controversial subject, then this book will help you navigate towards the answers you are looking for. It is a must-read for all who are reflecting on the ultimate question of origin as this book is bound to shake the worldview of many and I enthusiastically recommend this book.

> Fr. Dr. Michael Amaladoss S.J., Theologian, author of 35 books,
> Director of the Institute of Dialogue with Cultures and Religions (retd.),
> Loyola College, Chennai, India.

Can you believe both science and the Bible? In this fascinating book on origins, Dr. Christie says that you can and presents a compelling case that harmonization of science and faith is possible through the fascinating Revelation Model. With clarity and skill honed over many years, she has brilliantly and methodically presented a masterful collection of evidence from various fields of science, including cosmology, biology, and geology, and argues that supernatural creation makes more sense than random chance. This book is a vital resource for parents, teachers, and pastors to approach the question of origin with confidence, and I strongly recommend this book.

> **Rev. Dr. D. Mohan, Vice Chairman, World Assemblies of God, General Suptd. All India Assemblies of God Churches, Founder, New Life AG church, Chennai, India.**

I am very happy to endorse the book—*Beyond the Boundaries of Science* by Dr. Latha Christie, who holds a Ph.D. in Engineering with a senior Engineering position at work. Most engineers would have more than enough satisfaction in their life with such achievements. But due to a revelation from God, she is connected to her spiritual side of the personality, and this book of hers connects Engineering/Science and Religion, which some people might see as contradictions. She explores the mysteries of our origins that lie around and beyond the limits of what science can observe and presents a thoughtful treatment of many vitally important subjects, like the beginning of the universe and life. If you are interested to find a connection between God and science, then I encourage you to read this book as it provides stimulating analysis along with a sense of wonder.

> **Moshe Panski, President (retd.), A High Tech Co., Israel**

Humankind has always been intrigued by the mystery of the origins of the universe and of life itself. Brilliant scientific minds have been investigating the nature of the macrocosm and the microcosm to try and remove the shroud surrounding this mystery. Scientific breakthroughs for gaining answers to the origin of life and the universe continue to be made, which lead to our improved understanding of the DNA (as the blueprint of life) at the cellular level, as well as of subatomic structures at very small scales, to the behavior of galaxies at cosmic scales. However, many questions still remain, including the most fundamental ones of who we are; why we are here; and what our relationship to the universe is. Dr. Latha Christie attempts to answer, in depth, the latter questions on the basis of biblical theology while showing that the answer is not at odds with modern scientific views on the origins of our multi-scale universe, but that science actually supports the existence of an eternal supreme intelligence, or creator/God, as the source of all. The book is very well written, filled with an extensive overview of modern scientific theories on the origins of the universe, and most of all, it is strongly recommended to anyone looking for answers to the fundamental questions of life and its connection to our mysterious universe.

Dr. Prabhakar Pathak, Professor Emeritus, ECE dept.,
The Ohio State University, US.

Every thinking person must read this book. It is thoughtful, lucid, and bold. In this well-documented book, Dr. Christie argues that both science and Scripture reveal the same Creator. She addresses this long-standing controversy with clarity using up-to-date scientific data. I endorse this wonderful book as I am sure that every reader will become better informed and better able to piece together the truth about our origins.

Rt. Rev. Dr. P.K. Samuel, Bishop, CSI Karnataka Central
Diocese, India.

Dr. Christie offers an intriguing approach in her book—*Beyond the Boundaries* for interpreting the Genesis creation account. She proposes a "Revelation Model" which both honors the book's literary and cultural context and recognizes the divine inspiration behind the description of the progression of creation. She argues that the days of Genesis coincide with the physical progression of the universe's and earth's history as revealed by the discoveries of modern science. She also beautifully describes how the details of creation demonstrate clear evidence for intelligent design. Dr. Latha Christie offers an important contribution to the study of the interaction between faith and science, and her book will prove an enjoyable and informative read for both scholars and laypeople alike.

Dr. Stephen C. Meyer, Director, Discovery Institute's Center for Science and
Culture, Seattle, USA, and the author of *Signature in the Cell*
and *Darwin's Doubt*.

Dr. Christie explains many novel concepts of science including basic physics, cosmology, and biology in her book—*Beyond the Boundaries of Science*. With deep insights and wide perspective, she has well described all the topics, without using any equations. She also shows several mysterious problems regarding the origin of the universe and life so that this book was finished off to be interesting and enlightening. The latter half of this book treats the relationship between science and religion, especially Christianity, which describes God as the all mighty Being full of wisdom. Science and religion have been supplementary each other, and may finally reach the same understanding of Universe and Life. I am happy to endorse this wonderful book.

    Dr. Tadashi Takano, Professor Emeritus, Institute of Space and Astronautical Science and former Professor, University of Tokyo, Japan.

The origin of the universe, the origin of life, the origin of the earth and moon and other heavenly bodies, including black holes and dark matter and its counterpart dark energy, are engaging the attention of scientists all over the world. At the same time, theology has been engaged in unraveling the mysteries of these aspects, attributing them to the power of omnipotent and omnipresent God. In this book, the author Dr. Latha Christie, herself being an accomplished scientist and engineer, has succeeded in summarizing the experimental evidence and the competing theories, such as a universe or a multiverse, to help the reader weigh up these ideas. As Sir Isaac Newton stated that he was standing on the shoulders of giants while formulating his theory of gravitation, Dr. Christie will engage the passionate, both young and old, to see further in resolving these fascinating problems of our existence in the universe. I wish the book a wide circle of readership.

    Dr. R.M. Vasagam, Space Scientist & former Project Director APPLE, ISRO, Former VC, Anna University & Chancellor, Dr. MGR Educational & Research Institute, Padmashri, Govt. of India.

Beyond the Boundaries of Science
Exploring the Cosmic Story

LATHA CHRISTIE

RESOURCE *Publications* • Eugene, Oregon

BEYOND THE BOUNDARIES OF SCIENCE
Exploring the Cosmic Story

Copyright © 2019 Latha Christie. All rights reserved. Except for brief quotations in critical publications or reviews, no part of this book may be reproduced in any manner without prior written permission from the publisher. Write: Permissions, Wipf and Stock Publishers, 199 W. 8th Ave., Suite 3, Eugene, OR 97401.

Resource Publications
An Imprint of Wipf and Stock Publishers
199 W. 8th Ave., Suite 3
Eugene, OR 97401

www.wipfandstock.com

PAPERBACK ISBN: 978-1-5326-8150-9
HARDCOVER ISBN: 978-1-5326-8151-6
EBOOK ISBN: 978-1-5326-8152-3

Manufactured in the U.S.A.

CONTENTS

PART 1
PROLOGUE — 1

1. EXAMINING THE QUESTION OF ORIGIN — 3

PART 2
AN INQUIRY INTO THE MYSTERIES OF SCIENCE — 17

2. INVESTIGATING THE ORIGIN OF THE UNIVERSE — 19
3. INVESTIGATING THE ORIGIN OF LIFE — 51

PART 3
PROBING THE BACKBONE OF SCIENCE — 93

4. A SEARCH FOR TRUTH: ACROSS FIELDS OF STUDY — 95
5. A SYNTHESIS: SCIENCE AND BIBLICAL THEOLOGY — 105

PART 4
AN INQUIRY INTO THE THE MYSTERIES OF SCRIPTURE — 115

6. WHEN FAITH AND REASON INTERSECT — 117
7. THE REVELATION MODEL — 127
8. THE FASCINATING SEQUENCE — 133

PART 5
EPILOGUE — 165

9. A SHOWDOWN WITH TRUTH — 167

POSTSCRIPT — 187

Bibliography — 191

*To my boys, Pradeep and Praveen whose love
facilitated this work*

Foreword

Humankind is excited to look at the Sun and the bright stars in the sky. From the ancient days, different theories have been put forth to explain the origin of celestial bodies and human life. Recently, there is a consensus among scientists that the Universe came into existence 13.7 billion years ago, from a glow and thereafter evolved into stars and galaxies. Our galaxy—the Milky Way—has many stars. One such star is the Sun, which has completed five billion years with planets including our earth. The earth has been the home to many living beings that can move or swim or crawl or fly including the superior creation, the human race, with an amazing brain and a powerful mind. Still, the most important issue before us is the evidence of the origin of human life. We are getting more and more data from telescopes, particle accelerators, deep space missions, and genome sequencers. Attempts are being considered to colonize Mars and to resettle humankind in the habitable planets beyond our solar system if they exist. Despite all our development of technology and ideas, despite all our discoveries and theories, there is much about our origins that remain mysterious. Therefore, there is a need for a comprehensive analysis of many of the cosmic puzzles that accompany our greatest scientific advances.

As a scientist for more than three decades, Dr. Latha Christie has carried out thorough research on this question of the origin and documented the evidence to make sense of the various mysteries and competing ideas, in this book—Beyond the Boundaries of Science. She has also analyzed what physicists are saying about the origin of the universe, and what biologists are saying about the origin of life, and has integrated various philosophical thoughts to make connections to human life.

Dr. Christie had tried to gather all available evidence meticulously to find out what happened before humanity even existed using the latest scientific findings in various fields like cosmology, archaeology, anthropology, biology, and geology. She has reviewed both the scientific data and theories and the various interpretations of the Bible's account of our origins. In this way, she has presented a compelling summary that points towards the conclusion that there is a supernatural Creator behind the cosmic events of the creation of the universe and the creation of life and suggests that harmonization of science and faith is possible.

This book is a scholarly, well-researched treatment of the origin of the universe and the origin of life, from the grandeur and beauty of the cosmos to the tiniest sub-atomic particles. I am sure that the evidence that she has gathered for this book will help curious readers to enhance their wisdom and understanding about the origin of the universe and human life. I commend this brilliant book for your reading.

Dr. A. Sivathanu Pillai, Dr. Kothari Chair professor,
Distinguished Scientist & Former Chief Controller (R&D), DRDO,
Former Distinguished Prof. ISRO, Founder BrahMos Aerospace,
Padmabhusan and Padmashree, Govt. of India.

PART 1
PROLOGUE

CHAPTER 1

EXAMINING THE QUESTION OF ORIGIN

"Two things inspire me to awe—the starry heavens above and the moral universe within.

Immanuel Kant

THE INTRIGUING QUESTION OF THE ORIGIN

There are certain events that get seared into the mind at an early age. You can never wipe out certain memories. A violent storm, a damaging fire, a lightning strike that brought tragedy or simply the starry sky on a dark night: all these, collectively, tell you that you are puny and helpless compared to these mighty forces, under the power of someone who runs the universe. When I look back to my childhood, I remember an occasion in 1969, when the whole world was celebrating the greatest moment in the history of space exploration—the moment Neil Armstrong took what is referred to a giant leap for mankind. He had just become the first person to walk on the moon. I was only four. With the awe and wonder of a child, I was listening to that story as my mother was feeding me. She also pointed to the sky and warned me that I should never tell a lie, as there is a God up there watching over me. Both God and the skies lit a spark of restless curiosity and a sense of mystery within me.

I grew up learning to spot the constellations and memorizing their names, along with my siblings, using the big books from my dad's library. Many of my memories are somehow linked to the skies. I often find myself gazing into the night sky, and I am amazed at the vastness of the universe. Just imagine this: There are more than 100 billion galaxies, each of which contains 100 billion stars! Grappling with the idea of billions of galaxies still amazes me. If you ever travel through the atmosphere as far as 100 kilometers (or 60 miles) above the surface of the earth, you will have found yourself in space. Why does space trigger so much curiosity? Probably because space doesn't seem to have an edge and it is the closest thing that we can associate with the infinite. It's no wonder we spend billions of dollars every year on space exploration. After the first walk on the moon, we sent the first space shuttle into space in the 1970s. That was followed by the first space station that sits in space, where it produces hundreds of kilowatts of power and two to four astronauts are in it at any one time. From Yuri Gagarin, in April 1961, until today, about 550 people have traveled into space.

On 14th February (Valentine's Day) 1990, the Voyager 1 spacecraft found itself 40 astronomical units (150 million kilometers) away from the sun, even beyond Neptune. From beyond our solar system, it looked back and provided images of our planet from further away than ever before. The earth looked like a "pale blue dot." This picture still inspires wonder in the viewer. Carl Sagan, a member of the Voyager imaging team, writes, "That's here. That's home. That's us. On it everyone you love, everyone you know, everyone you ever heard of, every human being who ever was, lived out their lives."[1] When we see how minute we are in contrast to the vastness of the entire universe, we cannot help but be humbled. Our thinking that we are the center of the universe becomes ludicrous.

When a group of astrophysicists attempted the impossible, to calculate the span of the universe, they discovered that the visible universe seems to stretch out at least 24 gigaparsecs in all directions. That's a radius of 78 billion light-years (1 parsec is a distance of 3.26 light-years). A beam of light can go around the earth seven times in one second. But, even though light travels so fast (186,000 miles per second), it takes about 100,000 years to traverse from one end of our Milky Way galaxy to the other. Now try to wrap your mind around how vast our universe is. This immensity explains why, on the way home on the Apollo 11, Neil Armstrong said, "It suddenly struck me

---

1   Sagan, *Pale Blue Dot,* 12.

that that tiny pea, pretty and blue, was the Earth. I put up my thumb and shut one eye, and my thumb blotted out the planet Earth. I didn't feel like a giant. I felt very, very small."[2]

The vastness of the universe fascinated me and instilled a feeling of transcendence—that there is more to the universe than meets the eye and drove me to physics and also to other branches of science, to explore the big and deep question about the nature of the universe and its origin: How did this whole universe come into being? Over time, I found that this question of origin has enchanted not only me but also many philosophers and scientists over many centuries. This is because a well-grounded understanding of this key question of origin—where did this entire universe come from? How did life emerge?—will help in exploring the other most profound and eternally significant issues: What is the meaning of life? And where do we go when we die?

THE HISTORY OF IDEAS ABOUT THE COSMOS

From ancient times, the skies were, for the longest time, regarded as the locus of divinity and were a source of fascination. The early travelers and sailors used the sun by day and the stars and moon by night to find their way. In today's postmodern culture of smartphones and GPS, knowledge of the constellations is rare, but in the ancient past, it was a necessity for survival. With the birth of philosophy around 600 BC, new lines of thought emerged in the Eastern Mediterranean, in the field of astronomy—a departure from the myths and whims of gods to various models of a starry spherical universe. This era saw extensive developments in mathematics (Euclid and Pythagoras), physics (Archimedes) and, mostly, the birth of universities to promote research and to investigate ideas (Socrates' and Plato's Academy). Through his electrifying writing, Plato powerfully explained his understanding of celestial matters by the *Allegory of the Cave*.[3] Speaking through Socrates, he says that we, like prisoners in a cave, can only see shadows of the appearances of real objects. To grasp the real fundamental nature of what is causing them, we need to free ourselves from our chains and leave the cave. While Plato considered events in the world to be just

---

2   McRae, "Neil Armstrong," *ASME*, July 2012.
3   Plato, *The Allegory of the Cave*, VII.

shadows cast by forms behind them, his famous student Aristotle appealed to observation or empiricism. Aristotle said that heavenly bodies were made of something different from the bodies which are on earth.[4] He believed in the geocentric universe, later described in more detail by Claudius Ptolemy, where the universe is spherical and finite, centered on circular earth—an idea which was still the most widely accepted during the early Middle Ages.

During the sixteenth century, a radical shift in the human perception of reality took place, and there was a departure from the existing geocentric model to a heliocentric model that depicted the sun as the center, with the earth and other planets revolving around it. Nicholas Copernicus, Tycho Brahe and Johannes Kepler were the three main men behind this shift in thinking, now known as "the Copernican Revolution." Copernicus, a priest, argued that the planets moved in concentric circles around the sun and that the earth, in addition to rotating about the sun, also rotated on its own axis and the apparent motion of the stars and planets was due to this combination. The publication of his book, *de revolutionibus orbium coelestium* (On the Revolutions of the Celestial Spheres), just before his death in 1543, triggered the Copernican Revolution and made a profound contribution to the development of science.

In 1609, Galileo built his first telescope and, from the mountains, made observations of the craters on the moon, of Jupiter's moons and of the stars, and these confirmed Copernicus' model of a heliocentric Solar System. In 1632, Galileo published his book Dialogue, in which he claimed that the Earth moves around the Sun. Because of the prevailing view of Geocentricity, he was put under house arrest. However, Heilbron in his book *The Sun in the Church* provides a magnificent correction to oversimplified accounts of the hostility between science and religion. He writes about the four Catholic churches that housed instruments which threw light on the disputed geometry of the solar system, serving as the best solar observatories in the world between 1650 and 1750.

This revolution started by Copernicus, Kepler, and Galileo, and

---

4   Aristotle, *On the Heavens*.

allowed us to understand why planets move as they do was finally completed by Sir Isaac Newton. As a pioneer in the development of calculus, he was one of the greatest mathematicians and one of the most influential scientists who ever lived. He offered groundbreaking, mathematical approaches and developed a mechanical view of the operations of nature in his *Principia Mathematica*. This describes the structure of the physical universe as operating according to particular, universal principles. The principles which describe the motion of objects on earth can also be shown to govern the movement of planets. The same gravitational force which attracted the apple to the earth could, in Newton's view, operate between the sun and the planets. His three laws of motion established the general principles relating to terrestrial motion, and they could be applied to celestial motion as well. A hundred years after Newton, Pierre-Simon Laplace, a French mathematician, derived from Newton's mechanics a closed, deterministic view of the world. This strict deterministic concept suggests that, if someone knew all the relevant facts about the state of a system at a particular time, the future of that system could be correctly predicted by intelligence, ruling out any possibility of external influence.

Beginning on the cusp of the twentieth century, the most profound shift in our understanding of the Cosmos occurred. On 15th November 1915, Albert Einstein used his imagination rather than mathematics to come up with his most famous and elegant equation that rules the universe, transforming our understanding of space and time—the general theory of relativity (GR). GR states that in space-time distortion of the geometry of the cosmos by massive objects produces the effect called gravity in the same way that a heavy sleeper distorts the geometry of a sagging mattress on which he sleeps. This makes all objects, from light beams (light gets bent around clusters of galaxies) to pebbles, to follow curved paths through space. Before this, Isaac Newton and other scientists thought that all matter strutted and paraded on the stage called space-time, but Einstein showed that it is no longer a static stage, that stage of space-time itself—dances folding and wrapping, stretching and growing or collapsing in response to the actors of energy and mass. "Spacetime grips mass, telling it how to move; and mass grips spacetime, telling it how to curve."[5] It is mind-boggling to know that space-time wraps itself around a star and disappears into a black hole. Because Einstein unveiled such a theory, that transformed our understanding of the cosmos, he secured a place for himself as one of the topmost scientific minds—causing Dr. Thomas Harvey to run off with Einstein's brain after its autopsy, because he believed that it had unusual brain anatomy.

But where does this fabric of space-time come from? To address this question of the origin of space-time, we need another theory that could explain the microscopic world and this theory is called Quantum Theory. General Relativity explains the universe in terms of things that are large and heavy, by focusing on gravity. In contrast, Quantum Mechanics explains how things happen at small, sub-atomic scales, by focusing on the other fundamental forces. Although classical physics expected to calculate what actually happens, quantum theory can only calculate the relative probability of various things that might happen. The Heisenberg Uncertainty Principle states that probabilities are intrinsic to physics. This was a death knell to determinism and also to naturalism. The deterministic view posited by classical Newtonian science dissolved into sets of probability waves in Hilbert space entangled particles and multi-dimensional, curved space-time.

> *the primary qualities of the physical world are mysterious—reality is veiled*

The theories of relativity and quantum mechanics revealed that the primary qualities of the physical world are mysterious—reality is veiled and all we can know is how things appear when we observe them. Immanuel Kant calls this "transcendental idealism." Many of the phenomena at the quantum level, like quantum entanglement and the double slit experiment, are awe-inspiring, suggesting that we are very akin to Plato's prisoners in the cave and still require a much deeper insight into the underlying ultimate reality.

---

5   Wheeler, *A Journey into Gravity*, 275.

THE MYSTERIES OF THE UNIVERSE

In this age of cosmology, astronomers are probing the forces that shaped our universe with new generations of giant telescopes on land and specialized instruments in space, like the Hubble Space Telescope. Hubble has completed its 27 years in space and is still sending back beautiful and revealing images, bringing the beauty of the cosmos into our lives. Such telescopes are documenting the behavior of the cosmos on the largest of scales, going back to the first moments of its existence and charting its evolution hundreds of billions of years into the future.

In this way, many discoveries became possible, such as the accelerating expansion of the universe, the dark energy, extra-solar planets, and white dwarf stars. These discoveries point to the enduring mystery of science and the many cosmic puzzles that accompany our greatest scientific advances. What started the Big Bang? How did the inflation come to an end? What is the mystery behind the dark energy that we assume is the reason for speeding up the expansion of the universe? These mysteries still remain unsolved because, at present, there is no scientific theory to explain the beginning condition of the universe. All attempts to reconcile General Relativity, that explains the macro world, and Quantum Physics, that explains the micro world, have failed.

Another mysterious occurrence, that doesn't have an agreed explanation, is the design of the universe, which seems to be finely tuned. Cosmology is producing more and more evidence that our universe is based upon numerous parameters that must be extremely fine-tuned for life to exist. If the strength of any of the four fundamental forces had varied only slightly, the universe might have flown apart and dissipated moments after it was born. The position of our solar system, between the two spiral arms of Milky Way galaxy, the mass of the sun, and the mass and position of the earth are such that life can spawn on our planet. Was the basic layout of physical laws a coincidence? Why are the makeup of the universe and the laws that govern its working fine-tuned to an amazing degree? This fine-tuning of the initial conditions appears to conflict with the idea suggested by some scientists that the universe is not purposeful, but simply random.

There are several mysteries surrounding the origin and variety of life. Even after many decades of intense research, the origin of the first biological organism—the spontaneous generation of life—still remains a mystery. Even after unearthing so many fossils, the evidence is not enough to explain how different species arose. The beautiful complexity of DNA, the double helix structure which provides the blueprint of an individual, is another mystery. Our understanding of biology and genetics is not able to answer how a complex system like our eye might have come into existence. Also, the mystery shrouding the emergence of intelligence and morality in human beings appears unfathomable.

THE BOUNDARIES OF SCIENTIFIC TRUTH

The profound mysteries of life and cosmos still remain, despite great scientific progress in the disciplines of cosmology, archaeology, anthropology, biology, and geology, which helps us to probe deeply into the past. Although the rapid pace of scientific advancement provides a powerful way of searching for truth regarding the question of origin, it also brings many uncertainties. This is because scientific knowledge is only a collection of verified explanations about objective reality that are based on predicted and observed phenomena and so, as the quality of our observations increases with technological advances, our knowledge also changes. Thus, scientific ideas are constantly being revised and, in that process, what was previously considered scientific truths have sometimes turned out to be scientific half-truths or even falsehood. For example, according to Newton, the events and contents of the universe do not affect time, but, according to Einstein, velocity, gravity, and even location do affect time—a radically different picture. Though the laws of physics are the same from the beginning, with the advent of quantum physics—it looked as though we had discovered something novel. It was the limit of scientific knowledge that made us assume that nature was deterministic earlier and probabilistic now, shifting between two opposing and distinct classes of principles and beliefs. In this way, as scientific method tests evidence, some theories are disproved so that our understanding can be improved. Indeed, every theory or law will have conditions in which it is valid and outside of which it may not be. For example, Maxwell's electromagnetic theory is valid for macroscopic scales but fails when applied at subatomic scales.

Similarly, Ohm's law is valid for most materials, but not for semiconductors. For a particle physicist or an electromagnetic student, the evidence to prove that a subatomic particle or microwave exists generally has to be indirect. The evidence for belief in black holes or microwaves is not based on whether one can see or feel, but on the way, scientists have reached a consensus on these subjects, and the combined evidence in favor of them is strong.

Sharp controversy often exists among scientists over many of the fundamental ideas at the very core of the scientific method—one famous example being the Einstein-Bohr debate. Full confidence in the accuracy and completeness of a scientific theory is never possible. Far from providing a finished product, science is a work in progress. David Hume's work on empiricism led him to the understanding that scientific 'truth' is not absolute truth nor can it be proven by empirical observations.[6] Kurt Gödel's first incompleteness theorem says that even mathematical systems are incomplete and can contain contradictions and inconsistency, so it becomes difficult to validate the empirical observations of scientists using mathematics.[7]

"Based, then, on the work of both Hume and Godel, the conclusion is inescapable that absolute truth cannot be confined within the bounds of logical (inductive) or mathematical (probabilistic) systems. At best, all that can be done with induction or mathematics is to apprehend a part of the larger truth that is out there; the systems being used are simply not robust enough to capture the entirety of this truth."[8]

A scientist may claim and aim for objectivity in their collection and interpretation of data, but their approach and, therefore, its outcome is never completely free of the subjective influences on their thinking and philosophical orientation. This is very much true when trying to reconstruct the past. Astronomers, looking out into space, and biologists, looking back at the emergence of the first life, are very much like historians; they are trying to construct historical accounts of one-time events that happened in the past using the fragmentary evidence that

---

6   Hume, Treatise of Human nature.
7   Gödel, "Some mathematical results," 1930b, 1931, and 1931a.
8   Brush, *The Limitations of Scientific Truth*, 71.

is available in the present. When scientists attempt to describe the past, they are subject to cultural constraints and limitations in the same way as historians, who describe the past. Their reconstructions of the past may easily be distorted by the cultural biases, fears, and expectations of the present. Brush says, "Our senses are governed by our minds, and too often we see what we want to see, hear what we want to hear. Once again, facts do not speak for themselves; they must be interpreted!"[9]

Indeed, two scientists who are examining the same evidence can arrive at radically opposite conclusions. For example, Francis Collins, looking at DNA, says that it gives him a glimpse of God's mind, whereas Richard Dawkins calls it selfish. Both the scientists are looking at the same object and arriving at totally different conclusions. Science is certainly one approach to understanding our world, but what are its limits, and is it the only way?

> *Francis Collins, looking at DNA, says that it gives him a glimpse of God's mind, whereas Richard Dawkins calls it selfish.*

In February 2002, Donald Rumsfeld, the US defense secretary was awarded the Plain English Campaign's premier *Foot In Mouth* trophy by Britain's Plain English Campaign, for his most baffling comment: "as we know, there are known knowns; there are things we know we know. We also know there are known unknowns; that is to say, we know there are some things we do not know. But there are also unknown unknowns—the ones we don't know we don't know."[10] Some of the mysteries associated with our origin are so huge that our search may only end up identifying "known unknowns"—things that, due to our limitations, we do not know, but where we can tell that there is a gap in our knowledge. A "known unknown" might be something for which we have only certain clues so that we would need more information to see the whole picture. Alternatively, it might be something for which we have inexhaustible information, so much that its complexity is beyond our comprehension.

---

9   Brush, *The Limitations*, 248.
10   Remarks at a NATO press briefing, 2002.

Concerning the question of the origin of the universe and life, there can also be "unknown unknowns"—things we do not even realize that we do not know them. For example, there is a severe spatial limitation in the knowledge of the physical universe; scientific knowledge is confined only to a region of the universe that is observable. Beyond that radius, countless galaxies may reside, but they will be beyond our visible range as they are receding from us faster than the speed of light, towards the edge of the universe.

Most of the time, while describing a reality, certainty is replaced by uncertainty in quantum mechanics. Brush says that "Quantum theory takes away the certainty that scientists cannot hope to discover the "real" world in infinite detail, not because there is any limit to their intellectual ingenuity or technical expertise, nor even because there are laws of physics preventing the attainment of perfect knowledge. The basis of the quantum theory is more revolutionary yet: it asserts that perfect objective knowledge of the world cannot be had because there is no objective world."[11] Bernard d'Espagnat, the theoretical physicist who won the Templeton Prize in 2009, says that the world we perceive is merely a shadow of the ultimate reality because the reality revealed by science offers only a "veiled" view of an underlying reality that science cannot access.[12] He argues that underlying empirical reality is a mysterious, non-conceptualisable "ultimate reality," not embedded in space and (presumably) not in time either and such an objective reality is forever veiled from human knowledge.

THE INVESTIGATION: IN SEARCH OF TRUTH

If science ends up with many such "known unknowns" and also "unknown unknowns," how can we find a credible answer to the profound questions about origins such as: How did the universe come into existence? Was the basic layout of physical laws a coincidence? Is life a fluke or a lucky roll of cosmic dice? Are we a cosmic mistake? Is the Universe somehow fine-tuned to allow life to arise and flourish throughout the cosmos? Did time itself have a beginning, or was our beginning somewhere in the middle of an infinite quantity of time?

---

11 Brush, *The Limitations*, 48.
12 "Science Cannot Fully," *Science Magazine*, Mar. 2009.

How do you explain the rise of complex, intelligent life? What does it all add up to? Can we know the truth behind our origins through science alone? Is there a supernatural Designer behind all that exists? Does this imply that the universe is imbued with purpose? These questions have been intriguing me for years. Arthur L. Schawlow (Professor of Physics at Stanford University, 1981 Nobel Prize in physics): "It seems to me that when confronted with the marvels of life and the universe, one must ask why and not just how."[13] The why question became my major pursuit and encouraged me to investigate this question of origin.

> Is life a fluke or a lucky roll of cosmic dice? Are we a cosmic mistake?

In this pursuit, I found that there are many unsolved mysteries in the scientific understanding of our origin and this is not only because of the many "known unknowns" encountered in the search for truth but also due to the many "unknown unknowns" that might take us by surprise as we learn more. Can these mysteries ever be unlocked, or will they remain a mystery forever? Can science alone give a credible answer to my question of origin, or is there more, beyond the reach of science, such as a super intelligence responsible for this mystery? With this objective of an unrelenting quest for truth, I tried to gather all available evidence to find out what happened before humanity even existed.

I have been searching for answers, devouring articles on cosmology, geology, biology, archaeology and every other –ology that could give me information on this subject. I have been journeying along with astronomers and galaxy hunters, peering through their telescopes using my mind and imagination, to find the missing pieces of the cosmic puzzles, as I toured through the strange reaches of the galaxies. Like the stargazers, I spent time looking up and examining billions and billions of galaxies and was enthralled at the serene beauty of the star-studded skies. Like a curious child, I was immersed in a gauzy haze of adventure, and my mind often came to a standstill, as I stood in awe and wonder at the beauty of the cosmos. In this book, I aim to present

---

13 Margenau and Varghese, *Cosmos, Bios, Theos*, 105.

what I have discovered in an intellectually satisfying manner, though the complete truth cannot be grasped or presented exhaustively. I have aimed to avoid exaggerated claims or unfounded criticisms, but rather to sort out fact from fancy, as we grapple together in finding the right pieces of evidence.

One answer to the question "Where did we come from?" is based on the idea that there is a Divine Architect who has created a finely tuned universe that is imbued with purpose and functions in a consistent manner expressive of the character and the wisdom of its designer. The other answer to the question of origin, which is often assumed, is random chance. If random chance is the answer, and if life arose by random chance and natural processes, life could just as easily be extinguished in the same way. Chuck Colson says, "Our choices are shaped by what we believe is real and true, right and wrong, good and beautiful. Our choices are shaped by our worldview."[14] Since people can interpret the same evidence in diverse ways, it is vital to choose the right worldview, as our worldview colors the way we see almost everything around us.

Many intellectuals end up in the middle ground of open possibilities. If you are reading this book with such a stand, then please be sure that, though the possibilities are innumerable, the cumulative evidence presented in this book will help your ideas converge. Though one could spend a whole life gathering different ideas, enjoying the uncertainty and ambiguity and reveling in possibility, it is surely necessary at some point to narrow down those possibilities and confront reality. Mankind has for ages been seeking a credible answer to that one profound question of origin because it is on this that the meaning of this life and the possibility of eternal life depends.

While on the way to unlocking profound mysteries about our universe, I was caught in awe and wonder, on how little I know and on how much there remains to be learned. The dim haze of mystery shrouding the marvels of the natural world around me added enchantment to my pursuit of truth. I hope this book captures you with the same sense of wonder and awe.

---

14   Colson and Pearcey, *How Now Shall We Live*, 13.

PART 2
AN INQUIRY INTO
THE MYSTERIES OF SCIENCE

CHAPTER 2

INVESTIGATING THE ORIGIN OF THE UNIVERSE

A SEARCH FOR SCIENTIFIC EVIDENCE

My endeavor in this book is to explore a credible answer to the question of origin that has haunted humanity for ages, cutting to the very heart of human existence. Like a detective trying to discover the truth, I have made use of various approaches to empirical research, including the latest scientific evidence. It has been years since I started on this journey of finding the mysteries behind the origin of the universe and life, and what a journey it has been. A truly rewarding one! In this chapter, I present my findings on purely scientific approaches to explaining the origins of the universe and the next chapter goes on to describe the scientific approaches to the origin of life. The findings have been presented in such a way that readers can weigh up the evidence based on the cumulative results.

A GRAND QUEST: THE SEARCH FOR A UNIFIED THEORY

For more than three centuries, there has been an ongoing endeavor by scientists to unify the various phenomena of nature within a single theoretical framework and a single set of descriptive equations. During the seventeenth century, Sir Isaac Newton unified celestial and terrestrial phenomena, when he could explain the elliptical orbits of the planets through the same physical laws of motion that apply on earth–the law of universal gravitation. James Clerk Maxwell, regarded as the greatest theoretical physicist of the nineteenth century, unified electricity, magnetism, and optics. In June 1865, he published his treatise on electricity and magnetism in the Philosophical Transactions of the Royal Society, which described how electricity and magnetism are intimately connected. The paper, based on the experimental results of Oersted and Faraday, showed in a spectacular manner how the direct connection between electric currents and magnetic fields creates electromagnetic waves that can propagate through any medium, including perfectly empty-space, at the speed of light. Then came the colossal leap in understanding that light is also an electromagnetic wave, uniting electricity, magnetism, and light.

During the early twentieth century, the most profound shift in our understanding of matter occurred. Till then, atoms were considered to be the smallest units of matter, and the very word atom derives from the Greek *atomos*, which means uncuttable. However, a series of discoveries in a few decades debunked the unchangeable, unsplittable atom. By crashing elementary particles in giant accelerators, physicists found that the atom, consisting of protons, neutrons, and electrons, is not indivisible, but made up of subatomic particles. Subatomic particles can be divided into fermions (leptons and Quarks) and bosons (force carrying particles). Quarks are the building blocks of neutrons and protons. Electrons belong to leptons. Bosons are force carrying particles, corresponding to the four fundamental forces of nature—Gravitational force, Weak Nuclear force, Electromagnetic force and Strong Nuclear force. Gluons carry the strong nuclear force, Z and W bosons carry the weak force, the photons carry the electromagnetic force and the gravitons should be the force carriers for gravity. Neutrinos are similar to electrons, but have no charge, have an extremely small mass, and are one of the most abundant particles in the universe. There are at least seventeen known fundamental particles—six quarks, six leptons and four bosons plus the Higgs boson. Since fundamental particles come in more than one flavor, and more than one type, there are many more fundamental particles. But no one is sure why there are so many elementary particles and what determines their number.

The Standard Model of particle physics incorporates all known elementary particles and describes three of the four known fundamental forces (the electromagnetic, weak, and strong nuclear force, except gravity). For years, physicists were in search of the Higgs boson, dubbed as the "God particle," that helps elementary particles to gain mass. Finally, on 4th July 2012, CERN's Large Hadron Collider in Geneva, Switzerland, confirmed the discovery of the Higgs boson and particle physicists where thrilled. But even now, it is still not clear how and when the Higgs boson appeared after the Big Bang and gave mass to the mass-less particles. The mass of the Higgs boson is another mystery; it is around 126 billion electron volts or 126 times the mass of a proton, and this is precisely the mass needed for the universe to remain poised on the brink of instability. Stephen Hawking says that the Higgs boson could one day be responsible for the destruction of the known universe, and he called that day the Higgs boson doomsday.15 "Specifically, the fate of the universe depends quite sensitively not only on the Higgs but also on the mass of the top quark, another fundamental particle whose mass is around 180 GeV. Why is there this metastability? Why is the universe just at this point?"16 Is there any profound truth behind this?

> *For years, physicists were in search of the Higgs boson, dubbed as the "God particle"*

In 1900, Max Planck found that the energy in light waves comes in discrete chunks called quanta and that light can act both as a particle and as a wave. Light is both an electromagnetic wave and when quantized, it is a particle called a photon. This puzzle, called wave-particle duality, seemed absurd because, while waves are continuous, particles are discrete. Louis de Broglie guessed that, if light can act like particles, then particles like electrons can also act like waves and this idea won him, Nobel Prize for Physics in 1929. In 1927, the German theoretical physicist Werner Heisenberg shook the scientific world with his famous *Uncertainty Principle,* which states that precise simultaneous measurement of two complementary values, such as the position and momentum of a subatomic particle, is impossible; the more precisely one value is measured, the more erroneous the other value will be.[17] All

---

15   Dickerson, "Stephen Hawking Says," *Live Science*, Sept. 2014.
16   Das, "How the Higgs Boson," *Scientific American*, March 2013.
17   Heisenberg, *The Physical Principle*, 20.

these ideas led to the development of a mathematically consistent and coherent theory called quantum theory by brilliant physicists such as Neils Bohr, Heisenberg, Schrodinger, Born, and Pauli. Quantum theory unified particles and forces. According to quantum mechanics, particles are waves and, according to Maxwell, waves come from fields and, according to Newton, fields come from forces. A deep unification can be seen in quantum fields, which are a manifestation of both forces, by which matter interacts, and matter, of which particles are composed.

The Mind-Boggling Theory of Relativity

On November 15, 1915, Albert Einstein used his imagination rather than mathematics to come up with his most famous and elegant equation that rules the universe, transforming our understanding of space and time—the general theory of Relativity (GR). GR states that gravity is produced by massive objects distorting the geometry of space itself, in the same way, that a heavy sleeper distorts the geometry of a sagging mattress on which he sleeps. This makes all objects, from light beams (light gets bent around clusters of galaxies) to pebbles, to follow curved paths through space. Before this, Isaac Newton and other scientists thought that all matter strutted and paraded on the stage called space-time, but Einstein showed that that stage itself dances—folding and wrapping, stretching and growing or collapsing. "Matter tells space-time how to curve, and space-time tells matter how to move."[18] It is mind-boggling to know that space-time wraps itself around a dead star and disappears into a black hole. Because Einstein unveiled such a theory, that transformed our understanding of the cosmos, he secured a place for himself as one of the leading scientific minds, causing Dr. Thomas Harvey to run off with his brain after its autopsy on the basis that it had unusual brain anatomy.

The theory of Relativity showed that space and time are intimately connected, that warping or stretching of space-time is the cause of all gravity in our universe and that this affects matter, space, time, and even light. Starting from Newton, mathematicians were unable to explain the strangeness in Mercury's orbit, the little kink that none of the other planets have. However, Einstein's theory beautifully explained

---

18   Ford and Wheeler, *Geons, Black Holes*, 235.

that little kink in Mercury's orbit by showing that the Sun's mass was distorting the space through which Mercury was moving. The orbits of other planets are not that much affected as they are further away from the Sun. The fascinating thing is that this theory predicts some weird phenomena like astronauts in spaceships aging slower than people who stay on earth, solid objects changing their shapes at high speeds and time running more slowly at low altitude and passing faster at higher elevations. This curious phenomenon of relativity:

> previously has been measured by comparing clocks on the Earth's surface and a high-flying rocket. Now, physicists at the National Institute of Standards and Technology (NIST) have measured this effect at a more down-to-earth scale of 33 centimeters, or about 1 foot, demonstrating, for instance, that you age faster when you stand a couple of steps higher on a staircase . . . The NIST researchers also observed another aspect of relativity—that time passes more slowly when you move faster at speeds comparable to a car traveling about 20 miles per hour . . . Scientists refer to this as the 'twin paradox,' in which a twin sibling who travels on a fast-moving rocket ship would return home younger than the other twin.[19]

Mind blowing?

The Theory of Everything

Well before Einstein, it was the dream of physicists to find a single equation, a single overarching framework that would explain all physical phenomena of nature. But there was a great problem with the dual rise of Einstein's theory of general relativity and the theory of quantum mechanics, both of which have been robustly verified by experiments. An ideal framework should unify all the four forces: the force of gravity that is described by General Relativity and the three quantum forces that govern particles—the strong nuclear force, the weak nuclear force, and the electromagnetic force. However, such a master theory or a "theory of everything," still remains as one of the greatest challenges in physics, because General Relativity and Quantum Mechanics are not able to get along, and this conflict has been brewing

---

19 "NIST Clock Experiment," Sept. 2010.

for more than a century, since the days of Einstein. Though physicists have spent decades attempting to reconcile the two, Einstein's theory is still at odds with Quantum Mechanics and the process of reconciliation seems like working out a marriage relationship between two polar opposite people. This is because General Relativity, on the one hand, beautifully accounts for gravity and hence the big things of the universe—planets, galaxies, etc. and, according to it, the world behaves in a precise, predictable way whether it is observed or not. But Quantum Mechanics, on the other hand, accounts for the other three forces and is wonderfully adept at describing small things, the behavior of matter and energy at subatomic scales, where nothing is predictable, and objects don't have precise positions until they are observed. The clash is genuine, and it is due to incompatible descriptions of reality.

At the heart of this clash lies a concept which is known as a "singularity"—a point with zero volume and infinite density. There are two kinds of singularity in the Universe: one is the cosmological singularity, or the Big Bang, and the other lies behind the event horizon of a black hole, which is a massive star that has collapsed in. Though initially, scientists were cautious about the idea of a singularity, in the late 1960s, Roger Penrose with Stephen Hawking proved the possibility of a singularity through their study on black holes which suggested that the gravitational collapse of a large, dying star resulted in a point of infinitely small volume with infinite density. The evidence for the existence of singularity is convincing, as it is based on the observations combined with mathematical underpinnings, like the general theory of relativity and the standard theories of fundamental particles. At such singularities, an infinitely small point is required by general relativity, but such points do not exist in quantum mechanics, where everything is more like a smear due to the wave-like nature. Researchers around the world are perplexed and frustrated at this incompatibility and working hard in developing a new theory in the quest for reconciling the two theories. They believe that if they could reconcile these two theories—Quantum Mechanics and General

> *if they could reconcile these two theories—Quantum Mechanics and General Relativity—then they could solve many puzzles*

Relativity—then they could solve many puzzles like: what is inside a black hole? What happened at the birth of our universe? And why is the universe accelerating?

Leading Candidates for the Theory of Everything

Many decades have passed, but the puzzle of combining quantum mechanics and the general theory of Relativity remains unsolved. This is because, with gravity, the space-time stage, on which all the particles act, bends and warps under the influence of the particles acting in it, and in turn this bending and warping redirects the particles' motions. Since there are too many possible configurations of both the interactions and the underlying space-time, the mathematical models become complicated and break down. Also, since gravity affects all forms of energy, the calculation becomes complicated involving a whole bunch of infinities, especially at scales on the order of Planck scales $\sim 10^{-35}$ m. In the last few decades, two leading candidates have emerged for such a master theory: string theory and loop quantum gravity.

String theory suggests that everything is made of tiny strings.[20] All the subatomic particles, like quarks, gluons, and electrons, are not particles, but are vibrating strings, 10^{-33} meters long, vibrating at different frequencies, depending on their size. A single string can be any one of the particles, depending on its vibrational frequency, and each mode of vibration of a string yields a different particle. However, the mathematics of string theory relies on extra-spatial dimensions, which is impossible to experience directly. According to the earliest forms of string theory, our universe has a total of 26 dimensions; but, based on the latest advanced versions, known as Superstring theories, it has just ten dimensions. String theories have two major challenges: they are un-testable, and there are too many variants of the theory, and anyone of them could be correct.

Loop quantum gravity is concerned with the quantum properties of space-time and states that space is not continuous, but broken up into tiny chunks on the order of the Planck scale ($\sim 10^{-35}$ m) that are connected by links.[21] Elementary particles are produced based on the

---

20   Greene, *The Elegant Universe*, 2010.
21   Kiefer, *Quantum Gravity*, 2012.

tangling of these links. The loop quantum gravity approach also has two major challenges: first, they are untestable, and second, they have difficulty predicting how smooth space emerges at larger scales.

Why do these theories keep stumbling and why does a theory of everything still seem to be far from reality? Perhaps this is simply because we do not yet know enough about these phenomena, which cannot be seen; we might simply be missing the basic truths. Stephen Hawking says that "If we do discover a complete theory . . . then we shall all, philosophers, scientists and just ordinary people, be able to take part in the discussion of why it is that we and the universe exist. If we find the answer to that, it would be the ultimate triumph of human reason—for then we would truly know the mind of God."[22] But it seems that the mind of God may be too complex and incomprehensible for the limited human mind to comprehend. Thus, though the latest scientific knowledge has shown the inner workings of the universe at the subatomic level, there are still gaping holes in our scientific knowledge.

THE EXOTIC WORLD OF QUANTUM MECHANICS

The captivating branch of science called quantum physics, with its central tenet of duality, that light is both a particle and a wave, has triggered many to challenge the deterministic approach and to consider the possibility of the supernatural. Particle physicist Chad Orzel states: "Quantum theory's effect on science goes beyond the merely practical—it forces physicists to grapple with issues of philosophy. Quantum physics places limits on what we can know about the universe and the properties of objects in it. Quantum mechanics even changes our understanding of what it means to make a measurement. It requires a complete rethinking of the nature of reality at the most fundamental level."[23]

"Reality has been a murky concept. The laws of quantum physics seem to suggest that particles spend much of their time in a ghostly state, lacking even basic properties such as a definite location and instead existing everywhere and nowhere at once. Only when a particle is measured does

---

22   Hawking, *History of time*, 193.
23   Orzel, *How to teach Quantum Physics*, 5.

it suddenly materialize, appearing to pick its position as if by a roll of the dice."[24] Sometimes the results are bizarre; a particle can tunnel through a barrier that it classically could not surmount, as though a ball could tunnel through a hill, even though it doesn't have sufficient energy to go over it, and this magical phenomenon is called quantum tunneling.

There are at present numerous interpretations of quantum mechanics. The main two candidates are the many worlds (or multiverse) theory and the Copenhagen interpretation, proposed by Neils Bohr, over which he had a very strong debate with Einstein. This asserts that a particle cannot be assumed to have specific properties, leading to the phenomenon of a particle existing in multiple locations at the same time, which is called superposition. Schrödinger explained this concept, with his notorious thought experiment of the cat. He imagined putting a live cat into a box along with a small amount of radioactive matter and a vial of poisonous gas. During the test period, if anyone of the radioactive atoms decays, a mechanism will smash the vial of gas and kill the cat. Without observing or inspecting the inside of the box, we cannot be sure whether one of the atoms has decayed and so—cannot be sure whether the vial was broken; whether the gas was released and whether the cat was killed. Due to quantum superposition, Shrödinger's cat can be both dead and alive, until we open the box and check on the cat, whether it is either dead or alive. That is, its state becomes defined only when we open the box.

The other major interpretation of quantum theory is the many-worlds interpretation of quantum mechanics, proposed by quantum physicist Hugh Everett.[25] It was devised to overcome the assertion in the Copenhagen interpretation that an unobserved phenomenon can exist in dual states, that is Shrödinger's cat can be both alive and dead. According to this interpretation, the cat can be said to have "branched" out into two different worlds: in one world it is alive, and in the other world it is dead. The timelines of our universe constantly branch off, creating distinct and coherent worlds, each experienced by a different version of us.

---

24 Wolchover, "Have We Been," *Quanta Magazine*, June 2014.
25 Witt and Graham, *The Many-Worlds Interpretation*, 2015.

The puzzling nature of quantum mechanics challenges our imagination about the nature of reality, pointing to the need for metaphysics.

The Double-Slit Experiment

Take the case of the double-slit experiment. This involves etching two small slits in a screen and firing a beam of particles (e.g., electrons or photons) through them. This was first set up in the early 1800s by Thomas Young, a child prodigy, to prove that light was a wave. This simple experiment helped Young to challenge the belief of most scientists during that period, that light was only a particle, and to demonstrate convincingly the wave nature of light, because light passing through the two slits produces an interference pattern, that is characteristic of waves. When the light was quantized and sent as photons carefully one by one through the slits, they got one of the craziest experimental results of physics. The interference pattern could be seen even if the photons were fired one at a time. It appeared as though the photons seem to know where they would land if they were in a wave; they seemed to know where the previous photon landed or where the future photons will land. The experiment became mind-boggling when physicists placed detectors at each slit to find out which slit each photon was passing through. The very act of observing the photon by the detector caused the interference pattern to disappear. This logic-defying double-slit experiment that cuts to the heart of the weirdness of quantum mechanics is regarded by Richard Feynman as the "central mystery" of the quantum world.

In the mysterious span between creation and detection, a particle exists as a fuzzy mixture of all possible locations, encompassing all possible paths, called a superposition of states. Only when that particle is detected or observed that a location and the path it took to get there are determined and, before that, it is meaningless to attempt to define the particle's properties. The detection need not done by a conscious observer, as perceived by a few writers, it can be just a measuring device or an apparatus.[26] The many realities that existed collapsed at once, by the act of observing and this is famously known as the observer's paradox.

---

26   Zukav, *The Dancing Wu Li Masters*, 33.

The weird Quantum Entanglement Phenomenon

This idea of a material universe that is well-defined only at the moment of measurement (a "peekaboo" universe) is at the heart of Bohr's Copenhagen Interpretation. Einstein was against this idea, and he insisted on objective reality (universe exists independent of the mind of the observer), as he thought that there must be a hidden physical reality underlying the strangeness of the quantum world. So Einstein, along with Doris Podolsky and Nathan Rosen (EPR), proposed a quantum scenario, a thought experiment called the EPR paradox to argue that quantum mechanics is not a complete theory because the description of reality which it can give us is not complete.[27] The EPR paradox introduced one of the most mysterious ideas, called quantum entanglement:

> the result of a measurement on one particle of an entangled quantum system can have an instantaneous effect on another particle, regardless of the distance of the two parts . . . If two photons, for example, become entangled –that is, they are allowed to interact initially so that they will subsequently be defined by a single wave function–then once they are separated, they will still share a wave function. So measuring one will determine the state of the other: for example, with a spin-zero entangled state, if one particle is measured to be in a spin-up state, the other is instantly forced to be in a spin-down state.[28]

This effect is instantaneous, even if they are separated by a very large distance; even if one is at the other end of the galaxy. It appears as though the fates of the two particles are bizarrely linked. Erwin Schrödinger says about this, "I consider [entanglement] not as one, but as the characteristic trait of quantum mechanics, the one that enforces its entire departure from classical lines of thought."[29]

This mysterious phenomenon, that an object can be modified without any obvious contact with another object, is known as "nonlocality." It seems to involve something traveling faster than the speed of light. This bothered Einstein a lot, and he dubbed it "spooky action at a distance," because relativity suggests that nothing can

---

27 Bell, "On the Einstein-Podolsky-Rosen Paradox," 195–200.
28 APS, "Einstein and the EPR Paradox," Nov. 2005.
29 Schrodinger, "Discussion of probability," 555–563.

travel faster than the speed of light. At that time, Einstein had no way of knowing that future experiments would show that entanglement exists. In 1964, an experiment was developed by John Bell, to demonstrate quantum entanglement. The process involved entangling particles, then separating them by a significant distance and then testing whether, even when they were physically separated, there was still a "spooky" connection between them. Since then, other scientists have developed increasingly sophisticated experiments to test this, and most of the reports confirm that quantum entanglement is a real effect. A few critics say that all the experiments carried out so far have too many loopholes. However, in 2015, a group of Dutch physicists claimed to have closed all these loopholes and to have shown that this idea, which is one of the most surprising and puzzling ideas in the universe, the spooky action, is real, through their "loophole-free Bell test."[30]

> According to the research paper, the physicists made two diamond traps to capture a single electron—one per diamond—and used superfast laser pulses to "entangle" the electrons at a close distance. Next, they separated the diamonds almost a mile apart on Delft University's campus in The Netherlands...For their experiment, the physicists painstakingly set up the diamonds and lasers in a way that made it possible to measure one pair of electrons at a time—getting rid of one loophole. Closing a second loophole, they set up the diamonds far enough apart that there was no way the electrons inside could interact other than by entanglement. Based on the measurements of 245 pairs of entangled electrons, the team confirmed that each electron really was exerting "spooky action" on its entangled partner; whenever they measured one electron, the other electron across campus instantly flipped.[31]

Some commentators consider this entanglement to be the greatest mystery and call it "science's strangest phenomenon." Though the above phenomena may look like science fiction stories, we should bear in mind that quantum theory has been precisely tested. One of the most significant quantum physicists, Neils Bohr reflects on this:

---

30 Hensen et al.,"Loophole-free Bell inequality," 682–686.
31 Dickerson, "This bizarre experiment," *business insider*, Oct. 2015.

"Everything we call real is made of things that cannot be regarded as real. If quantum mechanics hasn't profoundly shocked you, you haven't understood it yet."[32] Such metaphysical aspects of quantum physics reveal that the subatomic world, with its unperceivable phenomena, has drawn us into a mysterious realm of awesome wonder and design. It points us to a world that is both law-governed and yet essentially indeterminate, with its probabilistic nature leaving some room for chance and top-down influences.

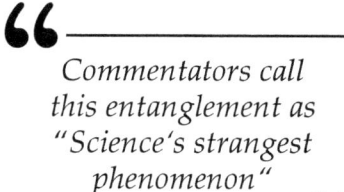
Commentators call this entanglement as "Science's strangest phenomenon"

THE ELUSIVE UNIVERSE

When Albert Einstein was formulating his theory of Relativity, he assumed a steady state universe that has the same gross properties and looks roughly the same in every direction, wherever an observer happens to be located in the universe. However, when Einstein tried to apply his General Theory of Relativity to the whole universe, he realized that his equations told him that the universe is not a static universe, but either an expanding or a contracting one. It appeared as if some mysterious antigravity force is acting on the universe. Ignoring it, Einstein inserted a counteracting force of repulsion into his equations, which he called the "cosmological constant," to prevent the universe from expanding or contracting. But his mathematics was better in telling the truth than he had wanted to believe. Within a little over a decade, Edwin Hubble peered into the sky using the powerful, 100-inch Hooker telescope and was astonished to find that the galaxies are indeed in motion and, by studying the red-shift of distant galaxies, Hubble concluded that our universe is expanding.

Einstein's Greatest Blunder!

Based on this discovery, Einstein denounced the idea of his cosmological constant as the greatest blunder of his life. But in the mid-1990s, many years after Einstein's death, when astronomers were trying to find out

---

32   Bohr, "Essays 1932-1957 on Atomic Physics and Human Knowledge."

whether the universe is expanding at the same rate or whether it is accelerating or slowing down, they found that the expansion is not slowing down but accelerating.

Research was carried out by two independent teams of astronomers, one based at the Lawrence Berkeley National Laboratory and the other at observatories in Baltimore and Australia, who used supernovas—giant explosions of a single dying star—as markers to measure the expansion speed at different times in the history of the universe. By measuring the speed and distance of many different supernovas, from many different eras, they found that things had changed over the billions of years of cosmic history, but in a way nobody expected. The expansion of the universe wasn't slowing down, but instead, it was speeding up so that there seems to be a mysterious antigravity force pushing it apart, called dark energy. They found that it was Einstein's statement, not the cosmological constant that may have been the true blunder. Einstein's "greatest blunder" was actually one of his greatest predictions.[33]

When Einstein's cosmological constant made a remarkable comeback in the name of dark energy, many experiments were conducted to investigate the mysteries of the dark energy, this antigravity that causes acceleration. Even with experiments that are 10 to 1,000 times more sensitive than the previous ones, this chameleon energy remains hidden from direct detection.[34] The Dark Energy Survey (DES) that uses a 570 megapixel Dark Energy camera, and the Wide-Field Infrared Survey Telescope, or WFIRST—are two of the current efforts to investigate the dark energy that is behind the accelerating expansion of the universe. Other than Einstein's cosmological constant, the other model of dark energy is quintessence. Scientists have not yet worked out which model is right. "If the quintessence theory is true and the amount of dark energy happens to increase, the story is even more dramatic. The expansion of the universe will eventually rip apart our galaxy, the solar system, Earth and eventually all matter so that individual atoms will be separated by unfathomable distances."[35] Thus

---

33   Lemonick, "Why Einstein was wrong," *LA Times*, Oct. 2011.
34   Redd, "Is dark energy caused," *Space*, Aug. 2015.
35   "Albert Einstein's colossal mistake," *Science News*, Nov. 2015.

the fate of the universe—whether the universe will run away from itself, and come to an end becoming dark and cold—seems to depend on the nature of this dark energy.

Is Dark Matter Dark?

Another mystery that scientists encountered while studying the cosmos is that that they have found that galaxies contain more mass than can be accounted for. This extra matter (around five times that of normal matter) prevents them from flying apart and also was involved in galaxy formation and in keeping them in stable galactic orbits. Scientists dubbed that extra matter dark matter, but it is still invisible and intangible, because it does not interact with anything, even light, but its presence is known via the gravitational pull it exerts.

> *the fate of the universe... seems to depend on the nature of this dark energy*

Thus the visible universe that we see around us—the earth, sun, stars, and galaxies is not only made of visible matter but also made of anti-baryonic, dark matter—and this became one of the most surprising discoveries of the twentieth century. "By fitting a theoretical model of the composition of the universe to the combined set of cosmological observations, scientists have come up with the composition: ~68 % dark energy, ~27 % dark matter, and ~5 % normal matter."[36] However, "much like dark energy, dark matter has remained an elusive material spotted only by its indirect effects on how regular matter behaves."[37] As "dark matter might be formed of weakly interacting massive particles, known as WIMPs,"[38] many experiments using the latest instruments have been hunting for this mysterious particle for decades, but it remains hidden still. Is dark matter just a phantom of scientists' minds? Can there be any other explanation?

Few researchers at the Weizmann Institute in Rehovot, Israel, has proposed an alternative theory to replace dark matter, called Modified Newtonian Dynamics or MOND, which could solve the problem of

---

36   NASA, "Dark Energy, Dark Matter."
37   Bruck, "Searching for dark matter," *Phys. Rev.*, Jan. 2018.
38   Gibney, "Dark-Matter Hunt Fails," *Nature*, Nov. 2017.

galaxy rotation curves.[39] According to MOND scientists, if dark matter existed, there would be more dwarf galaxies orbiting our Milky Way than the present. However, critics point out that MOND works well in individual galaxies, but not on clusters of galaxies.

While physicists continue the quest to detect WIMPs, "NASA's Hubble Space Telescope took an image of a bizarre, ghostly looking galaxy called NGC 1052-DF2 that astronomers calculate to have little to no dark matter. This is the first galaxy astronomers have discovered to be so lacking in dark matter . . . This unique galaxy contains at most $1/400^{th}$ the amount of dark matter that astronomers had expected."[40] For decades, astronomers believed that, without the dark matter, there could be no galaxies, and galaxies started their lives as bubbles of dark matter. But Hubble's discovery of DF2, with none of the dark matter, has shown that mystery of galaxies and everything we see around is much deeper.

The Mystery of the Great Attractor

The Hubble Space Telescope has revealed a mind-blowing glimpse into the vast stretches of our universe, and this makes us wonder whether we will ever be able to grasp the fullness of the cosmos. Galaxies are not spread out in the universe at random, but more packed together in groups, called superclusters, based on where in the universe they started to form. In the Observable Universe, there are around a massive 10 million superclusters made up of hundreds or thousands of galaxies. One of the largest known superclusters in the universe is the massive Sloan Great Wall that is around 1.4 billion light-years across and 60 times larger than the Milky Way. The Milky Way Galaxy that includes our solar system is part of the supercluster called Virgo. The observable universe is estimated to have a diameter of around 93 billion light-years.

The Milky Way Galaxy blocks our view of many distant galaxies and the region where astronomers find it difficult to look through is called the Zone of Avoidance. And deep in this Zone of Avoidance, there is a cluster of galaxies called the Giant Attractor.

---

39 Chown, "Forget dark matter," *New Scientist*, April 2014.
40 NASA, "Dark Matter Goes Missing."

Using CSIRO's Parkes radio telescope equipped with an innovative receiver, an international team of scientists was able to see through the stars and dust of the Milky Way, into a previously unexplored region of space. The discovery may help to explain the Great Attractor region, which appears to be drawing the Milky Way and hundreds of thousands of other galaxies towards it with a gravitational force equivalent to a million billion Suns ... our whole Milky Way is moving towards them at more than two million kilometers per hour.[41]

The universe as a whole appears to be very mysterious, and many like me are excited about the discoveries that are radically changing our present thinking.

THE MYSTERIOUS UNIVERSE

Though Science has solved many puzzles about nature and can even research the far galaxies and the minutest particles that make up the matter, there are still many missing pieces of the puzzle. This makes any researcher wonder whether there ever will be a Eureka moment. While there have been plenty of experiments and empirical data, we're still far from a convincing theory about the origin of the Cosmos. The universe remains as mysterious as it was in the beginning, for the early stargazers. Though there are many unsolved cosmic mysteries, two mysteries stand out—they have profound significance in our understanding of the universe and in this chapter we will look into the two cosmic puzzles in detail:

a. The beginning of the universe—how did our universe emerge?

b. The fine-tuning—why is the universe so finely tuned for life to exist?

The Beginning of the Universe

Anyone trying to figure out what happened during the first moments of the universe has to face a daunting challenge in physics. When did our universe emerge? Its age is estimated by scientists to be about 13.75 billion years. This is based on a theory called the Big Bang theory that

---

41   ICRAR, "Scientists discover hidden galaxies," Feb. 2016.

says that the universe that started 13.75 billion years ago will expand forever, becoming less dense with time. Virtually all cosmologists and theoretical physicists endorse this theory because of the overwhelming evidence for this theory.

Evidence for the Big Bang Theory

a. *Quasars*—the luminous galactic cores powered by black holes are objects from the past. Light from them takes several billion years to reach us, so they indicate that a few billion years ago the structure of the universe was very different from what it is today.

b. *Red Shift*—looking at distant galaxies, we find that the light has shifted down in frequency, toward the red end of the spectrum, revealing that they are moving fast away from us and that space is expanding.

c. *Abundance of chemical elements*—the existence of large amounts of deuterium, helium and other elements in our universe.

d. *Dawn's early light*—in the epoch of recombination, some 380,000 years after the Big Bang, ions of hydrogen and helium began to capture electrons and form atoms that were electrically neutral and allowing photons of light to cruise freely through the universe. The universe went from being opaque to being transparent, and this residual light from the Big Bang that permeates all of space is the cosmic microwave background radiation. The photons comprising this almost-uniform radiation background are the universe's oldest, having journeyed nonstop for nearly 14 billion years and they are now shifted down in frequency into the microwave range of the spectrum. Arno Penzias and Robert Wilson discovered this in 1965 when they noticed through their radio antenna that this diffuse radiation with a temperature of approximately 2.73 K emanated uniformly from all directions in the sky. The two missions by NASA, first the Cosmic Background Explorer (COBE) in 1992, and then the Wilkinson Microwave Anisotropy Probe (WMAP), in 2001 measured this microwave radiation with unprecedented precision. The satellite's measurements mapped the temperature variation—the primordial hot and cold spots in cosmic background radiation that formed the seeds of the giant clusters of galaxies and also pinpointed the age of our expanding universe.

The big problems of the Big Bang Theory

Though there is much evidence in support of the Big Bang theory, the theory faces two main problems:

a. Flatness Problem—why does the universe have such a "flat" geometry, with its density so nearly at the critical density? For the universe to be as flat as it is, even though curvature tends to increase with time, the conditions near the time of the Big Bang must have been fine-tuned to an unbelievable degree. Indeed, the universe would already have collapsed, if its density a billion years after the Big Bang had been only slightly greater, reaching the "critical density."

b. Horizon Problem—the microwave background radiation in all the regions of the universe looks the same in all directions though the regions are beyond each other's horizons. How can this be possible? If two regions should have the same temperature they should be close enough so that the information can be exchanged. Taking the speed of light (the fastest speed information can travel), if the time taken for the information to be exchanged between the regions that are isolated from one another (said to be beyond their horizons) is calculated, one can easily deduce that it needs a lot more time than is available. "The photons from the microwave background have been traveling nearly the age of the universe to reach us right now. Those photons have certainly not had the time to travel across the entire universe to the regions in the opposite direction from which they came. Yet when astronomers look in the opposite directions, they see that the microwave background looks the same to very high precision. How can the regions be so precisely the same if they are beyond each other's horizons?"[42]

The bang of the Big Bang

Due to the above problems, the Big Bang theory alone could not describe the initial bang; it could only explain the things after that

---

42 Astronomy Notes, "Embellishments on the Big Bang."

bang. To solve the problems with the Big Bang model the inflationary model was put forward in 1979 by Alan Guth. He calls his theory "as a theory of the 'bang' of the Big Bang as it describes the propulsion mechanism that drove the universe into the period of tremendous expansion," and he calls the universe as "the ultimate free lunch since it actually requires no energy to produce a universe."[43] Andrei Linde also proposed the eternally self-reproducing chaotic inflationary multiverse model. These inflationary models seemed to explain why the universe is flat and homogeneous. As per inflation, the universe underwent a period of rapidly accelerating expansion a few instants after the Big Bang that inflated the universe such that the observable part of the universe appears to be flat. Also, since the inflation pushed the inhomogeneities out beyond our event horizon, the homogeneous regions that appear to be separated now were in contact with each other before the inflation. The theory of inflation became popular as it seemed to solve the horizon problem and the flatness problem, though it faced the graceful exit problem—that there is no consistent way of explaining how the initial inflation slowed down to the lower rate of expansion that is prevalent now.

The theory of inflation also predicted the existence of primordial gravitational waves caused by the rapid expansion of the universe immediately after the Big Bang. Gravitational waves are ripples in the very fabric of space-time, produced by violent cosmic events such as black holes or neutron stars crashing into each other and merging. For the terrestrial observations of these waves, the major projects that have been established are: the BICEP2 (Background Imaging of Cosmic Extragalactic Polarization); the small aperture telescope at 150 GHz operated at the South Pole by Harvard-Smithsonian Centre for Astrophysics; the twin Laser Interferometer Gravitational-wave Observatory (LIGO) detectors operated from Livingston, Louisiana, and Hanford, Washington, USA; and the all-sky observations from space by the Planck mission launched by the European Space Agency. On October 3, 2017, three physicists—Rainer Weiss, Kip Thorne, and Barry Barish who pioneered LIGO, were awarded the Nobel Prize in physics for detecting the ripples of the gravitational waves

---

43   Bradt, "3 Questions: Alan Guth," *MIT News*, March 2014.

caused by the collision of two black holes that kinked the fabric of space-time a billion years ago.[44] However, this is very different from the primordial gravitational waves, produced by the early universe that seems to have a unique plasma motion, which produces a curling pattern of polarization, called a B-mode pattern. The detection of this B-mode pattern, the unique signature of the primordial waves, is being attempted by more than ten observational projects across the world as astronomers believe that this detection will help in solving one of the long investigated mysteries of the universe.

Few scientists claim that the BICEP2 experiment in the South Pole found the primordial B-mode polarization in the cosmic microwave background.[45] However, Steinhardt who was credited with inventing inflation along with Alan Guth and Andrei Linde says that their B-mode signal is due to known foregrounds: dust plus lensing plus synchrotron radiation plus systematics.[46] They have criticized the inflation theory arguing that it "cannot be evaluated using the scientific method."[47] They believe that the expected outcome of inflation (a smooth, flat universe with a certain spectrum of density fluctuations and gravitational waves) can easily change if we vary the initial conditions, as it is highly sensitive to the initial condition. Also, inflation generically produces a multiverse of outcomes—literally an infinite number of patches with an infinite diversity of possibilities, and there is currently no criterion to prefer one possibility over another. Individually and collectively, these features make inflation so flexible that no experiment can ever disprove or falsify it. However, Scientific American later published a letter by 33 scientists, including Stephen Hawking, dismissing the objections of Steinhardt et al. to inflation.[48]

Though the beginning condition of our star-studded universe is still a great mystery waiting to be solved, the beginning of the universe in a big bang some 14 billion years ago has been universally agreed scientifically. Professor Stephen Hawking puts it this way: "All the

---

44 Overbye, "LIGO Detects Fierce Collision," *NY Times,* Oct. 2017.
45 Moskowitz, "Gravity Waves from Big Bang," March 2014.
46 Horgan, "Physicist Slams Cosmic," *Scientific American,* Dec. 2014.
47 Ijjas, "Pop goes," *Scientific American,* Feb. 2017.
48 "A Cosmic Controversy," *Scientific American,* Feb. 2017.

evidence seems to indicate that the universe has not existed forever, but that it had a beginning, about 15 billion years ago."[49] It is no wonder that the Big Bang should be the most significant event in the history of the cosmos because all that exists started from that point.

The question, "Did the universe have a beginning?" has an emphatic answer, "Yes." In the realm of science, everything should have a cause. The Big Bang should have a cause. What caused it? How did the Universe emerge? Why was energy balled up in an infinitely pressurized mass? Why is there something rather than nothing?

The Cause of the Big Bang?

To find out what caused the Big Bang, or what happened near the initial singularity, scientists had to marry the two incompatible descriptions of reality—the General Theory of Relativity, because of dealing with the whole, the massive universe, and Quantum Mechanics, because it all began very small in a singularity. Since there is no single ultimate theory to explain the beginning condition, many scientists have put forward various theories to avoid the initial singularity and four such theories are analyzed here in detail:

a. Eternal Inflation and Oscillatory Universe

b. Universe out of Nothing

c. The Vilenkin's Tunneling Proposal

d. The Hartree Hawking No Boundary Proposal

Eternal Inflation and Oscillatory Universe

The oscillatory universe model, postulated to avoid the beginning, holds that our universe undergoes an endless succession of bangs and crunches following an expansion-contraction-expansion pattern, and the present universe bounced into expansion from a pre-existing universe that had been contracting.[50] However, Alexander Vilenkin states that "A cyclic universe runs into the second law of thermodynamics, which says that any system left to itself eventually reaches the state of maximum disorder, called thermal equilibrium.

---

49 Hawking, "*The Beginning of Time,*" Lecture, 1996.

50 Dunning, "The Big Bang might," *Imperial College*, July 2016.

So if the universe were cyclic, then in every cycle, the disorder in the universe would increase. Eventually, the universe would reach this thermal equilibrium state, which is a totally featureless mixture of everything—this is not what we see around us."[51]

Also, based on observations, scientists have found that the universe is not dense enough for gravity to stop and reverse its expansion.

> For decades, cosmologists were hoping that the invisible "dark matter" in the universe is enormous enough to accomplish a re-collapse of the universe. However, Robert Jastrow says that even if we assume that matter consists of 99 % of dark matter, it is still more than ten times too small to bring the expansion of the universe to a halt which indicates that the universe will expand forever. And in 1983 and 1984, American astrophysicists Marc Sher, Alan Guth and Sidney Bludman demonstrated that even if the universe contains enough mass to halt its current expansion, any ultimate collapse would end in a thud, not a bounce.[52]

Universe out of Nothing

Quantum creation from "nothing" involves the idea that the universe could have emerged from nothing, suggesting that the universe is like a long lived virtual particle (a short lived subatomic particle) that arose out of a random and extremely large fluctuation of energy in the quantum vacuum.[53] Rather than remaining perfectly smooth and continuous, if space and time are quantized, they can fluctuate, virtual particles can be created, and the universe can pop out of nothing. But what is nothing? If everything can be removed from the universe, there are still quantum fluctuations embedded in space-time and "then "nothing" is everything we see around us, and "everything" is nothing."[54] "If what we formerly took for nothing turns out, on closer examination, to have the makings of protons and neutrons and tables and chairs and planets and solar systems and galaxies and universes in it, then it wasn't nothing."[55]

---

51 Mitchell, "In the Beginning Was," *Tufts Now*, May 29, 2012.
52 Bonto, "Scientists Abandon the Oscillating Universe Theory."
53 Tryon, "Is the Universe," *Nature*, Dec. 1973.
54 Cain, "What is Nothing," *Universe Today*, August 2014.
55 Albert, "On the Origin of," *The NY Times*, March 2012.

It is clear that not only the term "nothing" is redefined here, but also that nothing, or the quantum fluctuations, must have been initially created somehow. The question is now pushed one step back: where do these vacuum fluctuations come from? This model suffers from a deep internal incoherence, because it is impossible to precisely specify, when and where in the primordial vacuum a fluctuation will occur, which will then grow into a universe, and there is surely a probability of multiple universes arising, which would collide and coalesce with one another.

The Vilenkin's Tunneling Proposal

On the other hand, Prof. Vilenkin postulates that in the beginning, the universe would have been a tiny, closed metastable spherical one, filled with a false vacuum that arose via a quantum tunneling event.[56] He says that this description for creating the universe from nothing has a nice mathematical description. However, this proposal "does not imply a large universe like the one we live in, but rather tiny curved universes that would collapse immediately. This is because of Heisenberg's uncertainty relation that predicts that not only smooth universes can tunnel out of nothing, but also irregular universes and the more irregular they are, the more likely they tunnel out."[57] Also, it is not clear how such a metastable early universe could have evolved into the stable, expanding universe we now see.

The Hartree Hawking No Boundary Proposal

According to this proposal, the very beginning of the universe occurred in imaginary time, before the (tiny) Planck time of the early universe. Hawking posits that, if we journey backward in time towards the beginning of the universe, time gives way to space near the initial moment such that we have space but no time.

> There wouldn't be any singularities in the imaginary time direction, at which the laws of physics would break down. And there wouldn't be any boundaries, to the imaginary spacetime, just as there aren't any boundaries to the surface of the Earth. This absence of boundaries means that the laws of physics

---

56 Mitchell, "In the Beginning," *Tufts Now*, May 2012.
57 "No Universe without Big Bang," *Physics*, June 2017.

would determine the state of the universe uniquely, in imaginary time. But if one knows the state of the universe in imaginary time, one can calculate the state of the universe in real time. One would still expect some sort of Big Bang singularity in real time. So real time would still have a beginning. But one wouldn't have to appeal to something outside the universe, to determine how the universe began. Instead, the way the universe started out at the Big Bang would be determined by the state of the universe in imaginary time.[58]

The universe would start at a single point, which wouldn't be a singularity, by transforming the conical hyper-surface (such as the tip of the cone) of classical space-time into a smooth, curved hyper-surface having no edge as given in the figure below.

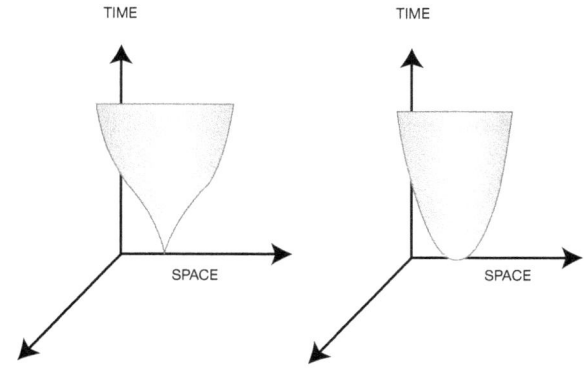

This mathematical idea manages to get rid of the singularity, suggesting that such a universe could quite literally be created out of nothing. Although imaginary numbers are useful tools in computations, it is very difficult to comprehend practically, as there's no real number that is the square root of a negative number. The use of imaginary numbers is just a mathematical device to make the equations easier to solve and, when one reconverts them to real numbers, the singularity, where the laws of physics breaks down, reappears again. Hence, this model is a grossly over-simplified model of the universe, and its mathematical exercises are highly speculative, with no empirical support and no verifiable scientific predictions.

---

58   Hawking, "The Beginning of Time," *lectures*.

Will we ever unlock the secrets?

The cosmologists have still not found what the universe was doing before it started expanding at the Big Bang. After all, the beginning of the Big Bang was not the real beginning. Though by refining their theories, the physicists could bring the early moments of the universe into sharper focus, the question remains—will we ever truly know what caused the Big Bang? Will it be ever possible to understand the universe, especially at about 10^{-36} seconds after the Big Bang, when the inflation started? It's a daunting challenge and the pace of progress is slow.

The Fine-Tuning of the Universe

The next question that haunts researchers is why the universe looks so orderly and predictable, working according to consistent, natural laws. Where do such laws come from? How come they are consistent and have elegant mathematical relationships? Why does the universe appear intelligible, so that the human mind can grasp it? The structure of the world, gravity, atmosphere, ecosystems, all of these are just right to support life, particularly human life on this planet. Everything here, from the light from the sun to the very laws of physics, suggests that the universe was carefully designed such that it was expecting life to come into existence. Various remarkable features of our universe, like the large size and low curvature of the present universe, the approximate homogeneity and isotropy of the matter distribution, seems to strongly suggest that reaching this present state is not random but highly special. It is startling that the properties of our universe are so exact that they allow life to thrive. Such an inquiry into the origin of the laws of nature and features of our universe points to the idea that it has been intelligently designed. Freeman Dyson says, "The more I examine the universe, and the details of its architecture, the more evidence I find that the Universe in some sense must have known we were coming."[59] The order and design of the universe suggest that some intelligence is behind it. British Astrophysicist Paul Davies says, "There is for me powerful evidence that there is something going on behind it all . . . It seems as though somebody has fine-tuned nature's numbers to make

---

59   Dyson, *Disturbing the Universe*, 250.

the Universe . . . The impression of design is overwhelming."[60] Even Fred Hoyle remarked: "A commonsense interpretation of the facts suggests that a super-intellect has monkeyed with physics, as well as with chemistry and biology."[61]

Cosmologists grapple with this apparent fine-tuning in the same way that they grapple with the beginning of the universe. They struggle to get hold of the mechanism that caused the fundamental constants, physical laws and the initial parameters of the universe to take on stringent, precise values so that intelligent life can exist in our universe. Why are there so many surprising coincidences? Why does the universe seem to have this amazing fine-tuning? Why do the constants have these values? Why they are so finely balanced? Hugh Ross gives an account of more than two dozen parameters that must have values falling within very narrow ranges if life is to exist.[62] Here are just a few of the many exquisitely fine-tuned parameters or cosmic coincidences:

a. *The strong nuclear force and carbon resonance:* "The formation of heavier elements, beginning with carbon, very sensitively depends on the balance of the strong and weak forces. If the strong force were slightly stronger or slightly weaker (by just 1% in either direction), there would be no carbon or any heavier elements anywhere in the universe, and thus no carbon-based life forms like us to ask why."[63]

b. *The strength of the force of gravity:* "If gravity, which is responsible for pulling the atoms together, into stars and galaxies had been a bit stronger, it would have pulled all the atoms together into one big ball and so stars and the galaxies would have been smaller, and more crowded. If its strength had been a bit weaker, the atoms would have distributed so widely that they would never have been gathered into stars and galaxies. If the force of gravity is increased by more than 3000 times than its actual value, any creatures near the size of human beings would be crushed."[64]

---

60   Davies, *Cosmic Blueprint*, 203.
61   https://en.wikipedia.org/wiki/Fred_Hoyle
62   Ross, *The Creator and the Cosmos*, 150-151.
63   Borwein and Bailey, "When science and," *Physics*, April 2014.
64   Collins, "The Fine-tuning Design," in Reason for the Hope.

c. ***The proton-to-electron mass and neutron-to-proton mass ratio:*** "The mass of a proton is roughly 1836.1526 times the mass of an electron. If the ratio is changed by any significant degree, many essential compounds for life, such as DNA, would not exist."[65] The mass of a neutron is a little more than the total mass of a proton, an electron, and a neutrino. If the neutron's mass were 1 percent less, then isolated protons would decay into neutrons, with the result that few elements heavier than lithium could form. On the other hand, if its mass were a bit larger, hydrogen would no longer fuse into deuterium and it would become impossible for starts to form.

d. ***The cosmological constant:*** This refers to the energy density of the vacuum, which was initially introduced by Einstein into his General Theory of Relativity. It is very small (even smaller than Planck's constant) and, " ... physicist Steven Weinberg in 1987, who argued from basic principles that the cosmological constant must be zero to within one part in roughly 10^{120}, or else the universe either would have dispersed too fast for stars and galaxies to have formed, or would have re-collapsed upon itself long ago."[66]

The Multiverse Hypothesis

Modern physics keeps making more advances so that these fundamental constants can be measured more and more precisely. Cosmologists and physicists are now confronted with this problem: Are all these many mysterious coincidences that make life possible, just a fluke? Has someone fixed it? How do we explain so many cosmic coincidences? Are they potential pointers to a super-intelligent Creator?

> *How do we explain so many cosmic coincidences?*

To circumvent this problem of fine-tuning, some cosmologists have assumed that our universe is part of a larger space-time of infinite volume—a multiverse that consists of very many universes. The laws of physics themselves vary between the different universes, so

---
65 Marsolek, *The Goldilocks Hypothesis*, May/June 2016.
66 Bailey, "What are the cosmic coincidences," March 2018.

only some of them are fine-tuned for life. If the constants vary from place to place and from time to time, then the right combination of constants that permit life to exist is bound to turn up somewhere, if there are enough universes. Since our universe is one among many, it is insignificant in the cosmic scale of things and we just happen to be in the right universe.

There are few theories that form the logical basis behind the multiverse hypothesis. The first one is the "many-worlds interpretation" of quantum mechanics, put forward to avoid the collapse of the wave-function or to solve the problem of Shrödinger's cat seeming to be both alive and dead, until it is observed. The next is the inflationary multiverse hypothesis by Alexander Vilenkin that states that the inflation is eternal and this eternal inflation creates an infinite number of bubble universes continuously and without end. "While some pockets of space stop inflating, other regions continue to inflate, giving rise to many isolated bubble universes."[67]

Then there is the multiverse theory, or M-theory, by Ed Witten, put forward to solve a series of embarrassing features of string theory.[68] M-theory added one extra dimension to string theory, making it 11 dimensions, and by that extended the one-dimensional strings into two-dimensional membranes and showed that the five varieties of the string theory could be reconciled in a single theory. The membranes must be on the order of Planck length 10^{-35} m. M-theory is not a single theory, but it offers 10 to the power of 500 logically consistent theories, each of which is capable of describing a universe. It shows many different ways of looking at the same thing. Max Tegmark believes that the multiverse is ultimately a mathematical structure, capable of adopting an infinite number of mathematical forms that correspond to different fundamental laws of physics.[69]

Though so many theories have been put forth to explain the multiverse model, all of them are un-testable, and astronomers are wondering whether they can be trusted.

---

67 Vilenkin, *Many Worlds in One*, 2006.
68 Witten, "Fivebranes and M-theory," *Nucl. Phys.*, 383.
69 Tegmark, "Is the Universe Made," *Scientific American*, Jan.2014.

Is there a Pre-existing Designer?

This idea of possible alternate worlds or universes, where everything turns out differently from ours, has been pretty entrenched in many science fiction movies and comics, such as the adventures of the Flash, Star Trek and even Buffy the Vampire Slayer, as well as lots of episodes of Doctor Who. The idea of possible alternate worlds illustrates the idea of contingency—a future event that is possible but cannot be predicted with certainty. For example, though it is easy to imagine a possible world in which we don't exist, it is not possible to tell with certainty. The existence of the other universes cannot be proved, as they are beyond our horizon and cannot be seen even if the technology improves and hence this is pure speculation. The multiverse argument to explain fine-tuning is similar to "the god of the gaps" fallacy because it works by inserting speculations into this gap in scientific understanding. Paul Davies says,

> Scientists are slowly waking up to an inconvenient truth—the universe looks suspiciously like a fix... Like Baby Bear's porridge in the story of Goldilocks, the universe seems to be just right for life. So what's going on?... The multiverse theory certainly cuts the ground from beneath intelligent design, but it falls short of a complete explanation of existence. For a start, there has to be a physical mechanism to make all those universes and allocate bylaws to them. This process demands its own laws, or meta-laws. Where do they come from? The problem has simply been shifted up a level from the laws of the universe to the meta-laws of the multiverse.[70]

Even if there are multiple universes, as this theory assumes, the mechanism behind the origin of these universes and their corresponding laws of physics is still unknown and, as Davies says, the problem of origin has simply been shifted up a level. Stephen Hawking and Thomas Hertog said, "A bottom-up approach to cosmology either requires one to postulate an initial state of the Universe that is carefully fine-tuned—as if prescribed by an outside agency—or it requires one to invoke the notion of eternal inflation, a mighty speculative notion to the generation of many different

---

70 Davies, "Yes, the universe looks," *The Guardian*, June 2007.

Universes, which prevents one from predicting what a typical observer would see."[71]

Hawking says, for the problems of the origin of the universe and the fine-tuning of its laws and constants, there are two options. To solve these two puzzles, one has to appeal either to a pre-existing designer or to a multitude of unseen universes and a set of untestable meta-laws.

CONCLUSION

As I describe this marvelous journey to the frontiers of the universe, I am now convinced that there remains a vast array of mysterious phenomena that are very difficult to explain, even after the audacious quests made by scientists who have stepped into the unknown, to ponder the unanswerable. There is a huge gulf in the scientific understanding of the origin of our universe.

In this chapter, I have presented some of the mystery surrounding the origins of the Cosmos, the gaps in our knowledge and the burning questions that baffle even the most colossal of intellects: how did the universe start? Why does the universe seem to be fine-tuned to allow life? Is there a 'theory of everything' which can make sense of both quantum physics, which we observe at small scales, and relativity, which we observe at large scales? These mysteries, made me ponder whether science alone, in principle, can explain everything, or are there any outside phenomena beyond this physical and observable reality?

Having considered the mysterious origin of the universe, let us go on, in the next chapter, to consider the mysterious origin of life.

---

[71] Hawking and Hertog, *"Populating the Landscape,"* Feb. 2006.

CHAPTER 3

INVESTIGATING THE ORIGIN OF LIFE

RECONSTRUCTION OF OUR PAST

According to a NASA report, till now around 3,500 exo-planets (planets orbiting other stars, beyond our solar system) have been found in more than 2,700 star systems.[72] In the last two decades, the search for alien life elsewhere has become one of the greatest quests. But still today, no trace of life has been found. The question—whether or not we are alone in the universe still seems impossible to answer. The existence of life seems to make our earth unique. Though we are obsessed about finding the possibility of life on other planets, the major question that has baffled scientists from the beginning remains unanswered—how did life originate on earth? A truth finder has to traverse through many mind-boggling discoveries in various fields, ranging from microbiology to geology to astrophysics, in his path to the answer to this question of the origin of life.

Recently a comment by India's minister for higher education, Satyapal Singh, stirred a hot debate on the Darwinian paradigm of evolution, with a commotion on social media and among scientists. He "suggested that Darwin's theory of evolution is 'scientifically wrong' and 'needs to change' in school and university curricula as 'nobody, including our ancestors, in written or oral, have said they saw an ape turning into a

> *The question—whether or not we are alone in the universe still seems impossible to answer*

---

72 NASA, "Our Living Planet Shapes," Nov. 2017.

man.'"[73] Articles in newspapers suggested that such hostility to science has resulted from loyalty to scriptures and that mythological accounts have marginalized hard evidence.

Time and time again, this argument between evolution and creation has rocked many societies and communities. But, when it comes to the origin and development of life, what evidence do we have?

Darwin's Theory of Evolution

Evolution, one of the most powerful ideas in science, was developed by Charles Darwin, born in England in 1809. Since Darwin's father was a physician, he too began his training to become a physician. However, that did not interest him, and so he moved on to Cambridge to study for the Anglican priesthood. After graduating from Cambridge in 1831, he went on a five-year journey as a naturalist to survey the South American coastlines and also visited the Galápagos Islands for five weeks. This journey helped him to collect thousands of important specimens and discover new species—finches, barnacles, beetles and much more. Upon returning to England, in 1837, while analyzing his discoveries, he began to suspect that one species can change into another. In the same year, he also sketched (in his notebook now known as the Transmutation Notebook B) the tree of life, to explain the interrelations among species based on their similarities and differences. He published his famous *On the Origin of Species* little more than two decades after he sketched this tree and set out to formulate his evolutionary ideas.

The tree of life paints a beautiful picture of the pathways by which life has progressed from unicellular organisms (protozoa or protists) to multicellular (metazoa), and then fish, amphibia, reptiles, birds, mammals and, finally, human beings. However, an article in the Guardian says that "Charles Darwin's "tree of life," which shows how species are related through evolutionary history, is wrong and needs to be replaced, according to leading scientist . . . We have no evidence at all that the tree of life is a reality."[74] These controversial claims incited

---

73 Bagla, "India's education minister," *Science Magazine*, Jan. 2018.
74 Sample, "Evolution: Charles Darwin," *The Guardian*, Jan. 2009.

curiosity in me to find out whether Darwinian evolution is a debatable topic because of religious beliefs, or because of the disagreements based on scientific evidence.

NEO-DARWINISM: THE PRESENT PARADIGM

Though Darwin published the Origin of Species in 1859, his theory faced many serious objections from biologists in the late nineteenth century due to the insufficiency of the fossil record and the limited knowledge of heredity. Here are some key points from a more detailed article—Darwin's theory states that natural selection works by favoring the most successful variants among the individuals in a population and it works based on the heritable cumulative changes that happen in some trait within a population over time.[75] But Darwin could not answer questions like: "Why do offspring vary? Why doesn't that variation get averaged away after a few generations? Natural selection may weed out poorer variants, but where do new variants come from?" Though Darwin's theory was not generally accepted at that time, after 1940 it was revived based on heredity experiments conducted in a monastery by Gregor Mendel, a Roman Catholic monk whose work had been contemporary with Darwin's. Mendel's theory of inheritance helped in the acceptance of Darwin's theory of evolution because it could explain why offspring resemble their parents, yet with plenty of variation among individuals from which natural selection can select the favorable ones and weed out the rest.

The prevailing paradigm of the theory of evolution uses Darwin's framework but also introduces Mendel's genetics, giving a theory known as "Neo-Darwinism" or simply "Darwinism." According to Darwinism, the history of life can be represented as a kind of family tree. Species that are alive today appear at the tips of the tree. A place where branches divide (a "node") identifies a species that was an ancestor to all the species higher up those branches. A node like this is called the "last common ancestor" of a group of organisms; and that group—all the tips that connect to the same node is called a "clade". The mechanisms behind the prevailing paradigm of evolution are two processes: mutation and recombination. Though the Darwinian claim

---

75   Owen, "Darwin's big problem," 2011.

is that there is good evidence for the overall shape of the history of life, with some informed guesses as to how life must have progressed, there is still tremendous dispute among scientists. The four main areas—where there is no scientific consensus will be explored in this chapter:

a. The origin of life on earth (Abiogenesis)

b. The origins of life's complexity

c. The transitional life forms

d. The reconstruction of the tree of life.

ABIOGENESIS—THE ORIGIN OF LIFE FROM NON-LIFE

Life's Rocky Start: Soup to Outer Space

During the middle ages, some people believed that life forms arose spontaneously from non-living matter because they noticed rats and flies coming out spontaneously from garbage, and meat and broths being covered with maggots and microorganisms when left exposed. This theory of the spontaneous origin of life was disproved in 1859, when the results of the study by Louis Pasteur, on the spontaneous generation of microorganisms in broths, were published. However, the origin of the first cell and the origin of replicating molecules have become topics of intense research by biologists for the past many decades. Famous universities like Harvard have funded the *origins of life* research project with millions of dollars.

While exploring the origin of life, scientists have made startling discoveries and they found a more sophisticated system of complexities inside the cell. In 1953, James Watson and Francis Crick discovered the genetic structure deep inside the nucleus of our cells called DNA—deoxyribonucleic acid, which has a double-helix structure. The DNA is packaged neatly into chromosomes and a human being has 46 chromosomes, 23 originating from the mother and 23 originating from the father. All the information in the DNA is stored as a code made of four chemical bases (nucleotides)—adenine (A), thymine (T), guanine (G) and cytosine (C). This discovery of deciphering the code of the DNA molecule has revealed something close to miraculous—it has an exquisite 'language' composed of more than 3 billion genetic letters, an information system that is more complex than anything that can ever be devised by the human mind. But who designed this? How the DNA molecule and the first cell came into existence? And how all these were assembled together so that the first life could begin? These are some of the scientific puzzles still waiting to be solved. Three main theories have been put forward to explain the beginning of life on earth:

a. Primordial Soup Theory

b. Hydrothermal Vent Theory

c. Panspermia Theory

Primordial Soup Theory

According to this theory, over time, life began from a series of chemical reactions when complex molecules, like amino acids and nucleotides that had accumulated in a warm environment came together and formed proteins and nucleic acids. The chemical reactions could have been triggered by some energy source like lightning strikes or ultraviolet light. This theory got its impetus when, in 1953, Stanley Miller and Harold Urey claimed to have shown through their experiment that the big leap from non-life to life is plausible, and biologically significant molecules can be formed by applying energy to mixtures of inorganic compounds. The chemical environment of the early earth was first reconstructed, and the energy to break down molecules in that simulated atmosphere was supplied by firing electrical sparks to simulate lightning in the water vapor and the gaseous mixture. Though the Miller and Urey's model showed a lot of promise initially, biologists found soon a host of problems with this model[76]—

a. The simulated atmospheric condition was radically different from the early earth's atmospheric condition. The experiment was simulated under reducing conditions with methane, ammonia, and hydrogen with no oxygen; however, there is much evidence that the early Earth had abundant oxygen. Some of the evidence can be summarized as: (i) Apollo 16 astronauts' studies reveal that early earth had oxygen as water breaks down into oxygen and hydrogen gas when bombarded by UV radiation. (ii)

---

76 Eastman and Missler, "The Origin of Life."

Oxygen is needed for ozone layer that is needed for the chemical building blocks of proteins, RNA and DNA. (iii) Geologists have discovered evidence of abundant oxygen content in the oldest rocks on earth.

b. The major outcome of the experiment was toxic capable of damaging the DNA.

c. The experiment produced only two significant amino acids found in living organisms (glycine and alanine), whereas a protein requires a chain made of more than 50 amino acids.

d. The experiment needs a chemical trap for the removal of the newly formed chemicals before the next spark, as the energy source (spark) can destroy the building blocks due to reversible chemical reactions. In the actual scenario, such a trap is not possible.

e. Due to reversible chemical reactions, preserving the chains of DNA and proteins, without intelligent guidance, is improbable even if there is a long period as the building blocks of DNA and proteins remains naturally un-bonded. So in a primordial soup that is moving towards equilibrium, there is no chance for the building blocks to remain bonded and, thus, for life to start.

f. In every life form on Earth, all sugars in DNA and RNA are always right-handed, and all amino acids in proteins are always left-handed. This bizarre, mysterious phenomenon is known as "homochirality," as it involves chiral ("handed") molecules. However, the Miller and Urey experiment, even after many decades of trying, produces only a mixture of 50 percent left- and 50 percent right-handed amino acids that cannot form enzymes with the correct three-dimensional shapes necessary for life.

Genes First or Metabolism First

"The problem of how a mixture of chemicals can spontaneously transform themselves into even a simple living organism remains one of the great outstanding challenges to science. Various primordial soup theories have been proposed in which chemical self-organization brings about the required level of complexity. Major conceptual obstacles remain, however, such as the emergence of the genetic code, and the "chicken-and-egg" problem concerning which came first: nucleic acids or proteins."[77]

Why this "chicken-and-egg" problem? For protein synthesis, the DNA is first transcripted into RNA and then relocated to the cytoplasm of the cell, where it is translated into protein. Every species creates daughter cells from a mother cell through a process called cell division or mitosis. During mitosis, a cell duplicates and splits the DNA (the chromosomes) to form two identical daughter cells. The DNA duplication requires numerous enzymes, which are proteins. While proteins are needed for DNA synthesis, DNA is needed for protein synthesis, and this creates the "chicken-and-egg" situation. This is one of the major hurdles for the idea of the chemical evolution of life. To circumvent this problem, two solutions have been proposed—genes first model and metabolism first model.[78]

a. *Genes First Model*

According to this model, the primordial world could have been an "RNA World," where RNA molecules formed first (rather than DNA or protein), and then RNA developed the ability to replicate, giving rise to a host of such self-replicating molecules.[79] In this model, RNA preceded DNA, protein, and metabolism. This is a "genes first" model because, in this, the replicator-first arose which evolved gradually through the improvement of its autocatalytic replication activity.

However, many biologists believe that this RNA World model is implausible because RNA is too complex to arise before there was life.[80] The backbone of the RNA molecule is the sugar (Ribose), which would be very difficult to form in the early earth conditions. "Ribose and other sugars have surprisingly short half-lives for decomposition at neutral pH, making it very unlikely that sugars were available as pre-biotic reagents."[81] Francis Crick says, "At

---

77 Davies, "The origin of life," *Science Progress*, 17.
78 Saladino et al., "Genetics first," *ChemSoc Rev.*, June 2012.
79 Ricardo and Szostak, "Origin of Life," *Scientific American*, 54–61.
80 Bernhardt, "The RNA world hypothesis," *Biol Direct.*, July 2012.
81 Larralde et al., "Rates of," *Proc. Natl Acad Sci.*, 81, 58–60.

present, the gap from the primal 'soup' to the first RNA system capable of natural selection looks forbiddingly wide."[82]

b. *Metabolism First Model*
This model states that the metabolism function might have preceded the replicative function. "Before the emergence of replication processes (making exact copies), a metabolic system must have reproduced (making similar copies), progressively increasing the accuracy of its pathways before allowing a spin-off system to initiate replication."[83]

However, in response, a few researchers have shown that "metabolic systems cannot retain compositional information well enough to allow such systems to evolve toward some kind of metabolic pathway. So metabolism-first models for the origin of life cannot possibly be a realistic way for life to have begun."[84]

Hydrothermal Vent Theory

According to this theory, life may have begun from the porous deep-sea hydrothermal vents, particularly alkaline vents.

The theory goes: At the time of life's origin, the early ocean was acidic and filled with positively charged protons, while the deep-sea vents spewed out bitter alkaline fluid, which is rich in negatively charged hydroxide ions . . . The vents created furrowed rocky, iron- and sulfur-rich walls full of tiny pores that separated the warm alkaline vent fluid from the cooler, acidic seawater. The interface between the two created a natural charge gradient...That battery then powered the chemical transformation of carbon dioxide and hydrogen into simple carbon-based molecules such as amino acids or proteins. Eventually that gradient drove the creation of cellular membranes, complicated proteins and ribonucleic acid (RNA), a molecule similar to DNA.[85]

---

82 Gesteland and Atkins, "The RNA World," 1993.
83 Danchin, "From chemical metabolism," June 2017.
84 Vasas et al., "Lack of evolvability," *PNAS*, Jan. 2010.
85 Ghose, "Origin of Life," *Live Science*, Jan. 2013.

However, many scientists disagree with this marine-origin theory, as the same process of proto-cell formation should happen today also in those alkaline hydrothermal vents—

> The earth's first cellular life probably arose on land, in vats of warm, slimy mud fed by volcanically heated steam, and not in primordial oceans. Such terrestrial environments boast the high ratios of potassium to sodium found in all living cells. Marine environments, meanwhile, are far too rich in sodium. For cells to synthesize proteins, they need a lot of potassium and not sodium as sodium blocks these activities. For life to evolve, proteins have to be synthesized for which potassium must be high. But in seawater, sodium outnumbers potassium 40 to 1. With this hurdle in mind, the team realized that the life could have originated on land not in oceans.[86]

Panspermia Theory

Panspermia theory suggests that life might not have begun on earth, but could have been brought here from space. Some scientists ascribe this to Martian meteorites that brought microbes here, as Mars could have had the chemicals essential to forming life, while early earth lacked some of them.[87] A few other scientists believe that life's ingredients were delivered to the Earth by asteroids, based on their findings of amino acids in some carbon-rich meteorites that came from a parent asteroid that entered an Earth-crossing orbit in 2008.[88]

> However, in order for life to originate elsewhere in the universe, there would have to be an environment on another planet capable of supporting it. Our study of the universe suggests that life as we know it would have a hard time surviving outside of the Earth. Even if extraterrestrial life did exist, proponents of the panspermia theory must still determine how life arrived on Earth. The best candidates to act as "seeds of life" are bacterial spores, which allow bacteria to remain in a dormant state in the absence of nutrients. In light of panspermia, the important question is whether bacteria or bacterial spores could survive in space. Since

---

86   Mosher, "Life on Earth," *National Geographic News*, Feb. 2012.
87   Kaufman, "Did Life on Earth," *National Geographic News*, Sept. 2013.
88   Fazekas, "Life Ingredients Found," *National Geography News*, Dec. 2010.

interplanetary distances are large, so the time a bacterial spore would have to spend in a meteorite or comet before hitting a host planet could range in the millions of years.⁸⁹

Even if spores could survive on an asteroid in space, they would also have to survive the heat of entering earth's atmosphere and the impact with the ground. This theory not only fails to identify the actual origin of life but also involves life traveling in ways which seem impossible.

Inferences

The origin of life is a profound mystery as even the best current models are just speculations. An article in Nature writes about Sir Fred Hoyle's view—

> The likelihood of the formation of life from inanimate matter is one to a number with 40,000 noughts after it . . . It is big enough to bury Darwin and the whole theory of Evolution . . . The chance that higher life forms might have emerged in this way is comparable with the chance that a tornado sweeping through a junk-yard might assemble a Boeing 747 from the materials therein . . . I am at a loss to understand biologists' widespread compulsion to deny what seems to me to be obvious.⁹⁰

The origin of life is a profound mystery as even the best current models are just speculations

Francis Crick, winner of the Nobel Prize in biology, says: "An honest man, armed with all the knowledge available to us now, could only state that in sum, the origin of life appears at the moment to be almost a miracle, so many are the conditions which would have had to have been satisfied to get it going."⁹¹

Stephen C Meyer in his book, *Signature in the Cell* takes up the question of how the immensely complex and exquisitely functional

---

89   Joshi, "Origin of Life," *Helix*, Dec 2008.
90   "Hoyle on Evolution," *Nature*, 105.
91   Davies, "The Cosmos," *Scientific American*, Sept. 2016.

chemical structure of DNA, which cannot be explained by natural selection because it makes natural selection possible, could have originated without an intentional cause. He argues that useful information comes only from an intelligent source and this argument for intelligent design is not based on ignorance or "giving up on science," but instead upon our growing scientific knowledge of the information stored in the cell. As per him, "Intelligence is the only known cause of complex functionally integrated information-processing systems."[92] He says, "Call it miracle, call it some other pejorative term, but the fact remains that the materialistic view is a truncated view of reality."[93]

The difficulties faced by researchers who are investigating the origin of life have been captured in a memorable anecdote:

> A physicist, a chemist, and a mathematician are stranded on a desert isle, when a can of food washes up on the beach. The three starving scientists suggest, in turn, how to open the can and ease their hunger. The physicist suggests they hurl it upon the rocks to split it open, but this fails. The chemist proposes they soak it in the sea and let the salt water eat away at the metal; again, no luck. They turn in desperation to the mathematician, who begins, 'Assume we have a can opener…' When discussing the evolution of life, biologists can often sound a bit like that mathematician. Beginning with a single cell, Darwinian evolution provides a simple, robust, and powerful algorithm for deriving all the astonishing richness of life, from bacteria to brains. Natural selection and other evolutionary forces, acting on surplus populations of replicating cells and multi-cellular organisms, lead inevitably to evolution and adaptation. Give biologists a cell, and they'll give you the world. But beyond assuming the first cell must have somehow come into existence, how do biologists explain its emergence from the pre-biotic world four billion years ago? The short answer is that they can't, yet.[94]

Is it possible for the biologists to give us the world once they get the first cell?

---

92 Meyer, *Signature in the Cell*, 346.
93 Chang, "In Explaining Life's," *NY Times*, Aug. 2005.
94 Robinson, "Jump-Starting a Cellular," NCBI, Nov. 2005.

THE ORIGINS OF LIFE'S COMPLEXITY

Charles Darwin, after proposing his theory, spent many years methodically compiling evidence for evolution as he anticipated many arguments and the greatest of all was how his method of gradualism could produce certain complex structures. Even a single living cell is complex. According to Paul Davies: "Living organisms are mysterious not for their complexity per se, but for their tightly specified complexity."[95] The oldest microorganism fossils found suggest that their genomes already included thousands of bits of carefully-arranged code and that they had immunity and defense systems capable of adapting to their environment. Strobel defines irreducible complexity as a system that "has a number of different components that all work together to accomplish the task of the system, and if you were to remove one of the components, the system would no longer function."[96] The wings of birds are irreducibly complex. Without a perfect design wings cannot function; for flight, wings need to be strongly attached, and themselves strong enough to support the bird in the air. The human eye consists of complicated parts such as the retina, lens and so on, which should be of the right size and shape to allow vision. For vision to occur at all, just for one light-sensitive part, many different proteins and systems would have to have developed and would have to be present together. "Darwin himself confessed that it was "absurd" to propose that the human eye evolved through spontaneous mutation and natural selection."[97] Many biologists think that the Darwinian mechanism of natural selection and random mutations can adequately account for such complexity.

> The human eye is made up of many parts—a retina, a lens, muscles, jelly, and so on—all of which must interact for sight to occur. Damage one part—detach the retina, for instance—and blindness can follow. In fact, the eye functions only if the parts are of the right size and shape to work with one another. If Darwin was right, then the complex eye had evolved from simple precursors. In On the Origin of Species, Darwin wrote

---

95   Davies, *The Fifth Miracle*, 112.
96   Strobel, *The Case for a Creator*, 197.
97   "Darwin's Greatest Challenge," *Science Daily*, Nov 2004.

that this idea 'seems, I freely confess, absurd in the highest possible degree.' But Darwin could nonetheless see a path to the evolution of complexity. In each generation, individuals varied in their traits. Some variations increased their survival and allowed them to have more offspring. Over generations those advantageous variations would become more common—would, in a word, be 'selected.' As new variations emerged and spread, they could gradually tinker with anatomy, producing complex structures.[98]

Paul Davies writes: "Natural selection . . . acts like a ratchet, locking in the advantageous errors and discarding the bad. Starting with the DNA of some primitive ancestor microbe, bit by bit, error by error, the increasingly lengthy instructions for building more complex organisms came to be constructed."[99]

However, some other scientists, such as Michael Behe and the other proponents of Intelligent Design (ID), strongly object the view that blind processes alone are the cause of such wonderful complexities. A research group in Seattle called the Discovery Institute, which is part of the intelligent design movement, claims that around 1000 scientists signed a petition stating their skepticism of Darwinism. Michael Behe argues that, since undirected evolution fails to explain such irreducibly complex systems as the optical precision of an eye, the little spinning motors that propel bacteria and the cascade of proteins that cause blood to clot, the evidence points to the hand of a higher being at work in the world.[100] While earthworms and sea urchins do not have eyes, caterpillars and starfish have simple flat eyespots. The human eye uses only one lens and retina but, in contrast, the eye of a fly has thousands of tiny columns and each one sees a tiny part of its overall field of vision. Vertebrates detect light with cells called ciliary photoreceptors, whereas the photoreceptors of insects and other invertebrates are rhabdomeric. The evolutionary path from the eye of a simple organism to the human eye cannot plausibly be explained by undirected evolution because the eyes of various creatures are totally different.

---

98 Zimmer, "The Surprising," *Scientific American*, Aug. 2013.
99 Davies, *The Fifth Miracle*, 42.
100 Behe, *Darwin's Black Box*, 1996.

Brian Miller says that the many supposed explanations of the evolution of complex features and the simulations describing how a mechanical eye could develop incrementally, scores high on imagination and flare but low on empirical evidence and thoughtful analysis.[101] This is because such explanations ignore the details of how a real eye functions and how it forms developmentally, leaving them disconnected from biological reality. He has presented a convincing argument, that for any species, upgrading to high-resolution vision requires massive reengineering in a single step and such radical innovation, coordinated to achieve a distant goal, is only possible with intelligent design. On the challenge of the minimum time required for hypothesized evolutionary transformations, such as the development of the camera eye, to occur through undirected processes, he says that even if the selective pressures were favorable, the required time scales are far longer for sufficient numbers of coordinated mutations to accumulate than the maximum time available, as determined by the fossil record. Thus the various breakthroughs in technology increasingly point to an intelligent Designer.[102]

> *the various breakthroughs in technology increasingly point to an intelligent Designer*

Inferences

Some features, such as wings and eyes, seem to be "irreducibly complex." In other words, it is not clear that a half-evolved version of them would have the survival benefit which is required for evolution to happen. The classic "March of Progress" image, a simple progression from a crouching ape to upright humans, depicted by Darwin's evolution, just by blind processes alone, cannot explain how the big, complex brains of humans could evolve from primitive bacteria.

"William Paley, an Anglican priest, made a similar argument in the early 19th century. Someone who finds a rock can easily imagine how wind and rain shaped it, he reasoned. But someone who finds a pocket

---

101 Miller, "Eye Evolution," *Evolution News*, Feb. 2017.
102 Miller, "Eye Evolution." Feb. 2017.

watch lying on the ground instantly knows that it was not formed by natural processes. With living organisms so much more complicated than watches, he wrote, 'The marks of design are too strong to be got over.'"[103]

MUSING THROUGH THE MISSING LINKS

Suppose there is a transition down through time, from one group of species to another group of species; a transitional fossil would be one that dates from the middle, between the two groups. A transitional fossil can give the details of how the transition happened. Let us assume that there is one group of species called A that has changed to species B that gave rise to species C, which in turn finally emerged as species D. Any fossils of species B and C are called transitional fossils. According to the theory of evolution, transition need not happen to all A's—it can happen to a particular group among species A, for example, to members that live in a specific geographic area, and so A can end up existing along with D.

Transitional forms should lie morphologically between the ancestral and descendant groups. Rather than being perfectly halfway between one group of organisms and another, a transitional form just needs to have aspects of an evolutionary change. Since every group of organisms is thought to have evolved from an earlier group, there should be an enormous amount of transitional forms between groups, but convincing examples of such transitional forms are very rare and fewer than evolutionary theory would predict.

Darwin said, about the transitional forms, that, "this, perhaps, is the most obvious and gravest objection which can be urged against my theory."[104] The absence of the necessary transitional forms, even one and half centuries after Darwin proposed the gradualistic model of evolution, has become a nagging issue plaguing Darwin's theory. "Darwin prophesied that future generations of paleontologists would fill in these gaps by diligent search. It has become abundantly clear that the fossil record will not confirm this part of Darwin's predictions.

---

103 Chang, "In Explaining Life's," *NY Times*, Aug. 2005.
104 Darwin, *The Origin of Species*, chapter IX,

Nor is the problem a miserably poor record. The fossil record simply shows that this prediction was wrong."[105]

Incompleteness of the Fossil Record

An organism becomes a fossil only under certain conditions. If it lies exposed and decomposes after its death, it cannot be fossilized. For an organism to become a fossil, it must not decompose or be eaten, but it should be buried soon before it decays. Also, "different species have different propensities to fossilize; the amount of rock fluctuates over geological timescales, as does the nature of the environments that it preserves."[106] The fossil record is very patchy and incomplete, making it difficult to work out what evolutionary changes happened and when. Most fossils preserve only teeth and bones, although a few show the features of soft tissues such as stained impressions of internal organs. Birds have light, hollow bones and so only a few fossils of birds have been found. Plants are problematic because their pollen, leaves, stem, and roots do not necessarily stay together even in life, let alone in the fossil record. Surprisingly, although rodents are the most common mammal today, there are hardly any fossils of them. Among apes, the orangutan is the only great ape with a good fossil record. Also, there is no certainty that a fossil has kept its true-to-life shape. The understanding gained from one single fossil keeps changing as the years go by.

The Mystery of the Cambrian Explosion

Cambrian Explosion happened during the early Cambrian period that began around 542 million years ago, at the start of the Palaeozoic Era during a period of profound environmental change combined with major extinctions. During this time, major animal body forms with mineralized skeletal remains suddenly appeared in the fossil record in a relatively short period in evolutionary history, around 20 million years or so. Biologists say that such rapid changes appear after a time of major extinction. The Chengjiang Maotianshan Shales in the Yunnan Province of China and Burgess Shale in Canada are some examples of superbly preserved fossils. The Cambrian Explosion

---

105 Eldredge, *The Myths of Human Evolution*, 45–46.
106 Donoghue and Yang, "The evolution of methods," 371.

presents challenging questions because it is at this time that the main branches of the tree of life should have emerged. But the major animal body plans that appeared at this time did not include modern animal groups—they were more primitive.

"Most of the animal groups that are represented in the fossil record first appear, 'fully formed' and identifiable as to their phylum, in the Cambrian, some 550 million years ago. These include such anatomically complex and distinctive types as trilobites, echinoderms, brachiopods, mollusks, and chordates. The fossil record is therefore of no help with respect to the origin and early diversification of the various animal phyla."[107]

The flowering plants or the "angiosperms appear rather suddenly in the fossil record . . . with no obvious ancestors for a period of 80–90 million years before their appearance."[108] That made Charles Darwin to write to a friend in 1879, that flowers were for him "an abominable mystery." Similarly, many mammalian groups seem to appear suddenly with no credible evidence of transitional forms.

Stephen Gould and Niles Eldredge advocated the theory of punctuated equilibrium to account for this lack of change recorded in the fossil record, suggesting that evolution happened in rapid bursts over short periods. This idea started ongoing arguments in evolutionary biology about whether divergence of species happened gradually or in sudden bursts? Is evolution punctuated or gradualistic? "This controversy, widely known as the 'punctuated equilibrium' debate, remained unresolved, largely owing to the difficulty of distinguishing biological species from fossil remains."[109] According to Michael Ruse, many paleontologists tend to favor punctuated equilibrium, while many geneticists hold to gradualism.[110]

The Cambrian explosion which puzzled Darwin poses many mysterious questions—Why such a sudden change in the evolutionary process? Why do species belonging to many main divisions of the

---

107   Barnes et al., *The Invertebrates*, 1993.
108   Bodt et al., "Genome duplication," *Trends in Ecology and Evolution*, 591-597.
109   Mattila and Bokma, "Extant mammal," *Proc. R. Soc. B: Biol. Sci.*, 2195.
110   Ruse and Travis, *Evolution: The First Four Billion Years*, 2009.

animal kingdom suddenly appear in the fossil record? Why are the precursors or prior ancestral forms to Cambrian fauna missing? And what sparked or triggered the Cambrian explosion?

> *The Cambrian explosion which puzzled Darwin poses many mysterious questions*

Regarding the Cambrian explosion, Stephen Meyer writes that building new types of organisms, in a geological blink of an eye, would require immense volumes of new biological information; and that intelligent design is the only known cause of the origin of large amounts of functionally specified digital information.[111] Such complex animal life that was springing into existence without precursors in the fossil record more than 500 million years ago, in the Cambrian explosion, points decisively to an intelligent cause.

A few Transitional Forms

The mystery behind the transitional forms is still growing with the findings of every new transitional form, triggering much controversy among scientists. Some transitional forms, which are the main sources of controversies, are discussed in this chapter.

Tiktaalik Roseae

Scientists accept that the evolutionary history of fish from jawless, bottom feeders to bony fish and then to amphibians and into land walking tetrapods is still not fully understood, as it is complex. The discovery of a 375 million-year-old fossil *Tiktaalik Roseae*, with its well-preserved pelvis and a partial pelvic fin, was initially thought to be a transitional fossil between fish and the four-legged animals called tetrapods.[112]

However, it lacked skeletal support structures and also lacked hind limbs, raising controversy and leading some scientists to regard it as a lobe-finned fish. Also, a discovery of tetrapod footprints in Poland, dated 397 million years, much before *Tiktaalik Roseae*, raises further

---

111 Meyer, *Darwin's Doubt*, 2014.
112 Dalton, "The fish that crawled," *Nature*, April 2006.

questions about the claim that *Tiktaalik Roseae* is a transitional fossil.[113]

Archaeopteryx

The fossil of *Archaeopteryx* has been the source of controversy from the time it was discovered. It was considered to be the missing link between birds and dinosaurs by some paleontologists, who suggest that "birds are living dinosaurs." But others vehemently oppose this, because they believe "birds are not dinosaurs." So the idea the birds are descended from theropods ("beast-footed" dinosaurs that walked on their hind legs) remains disputed. According to the proponents of "birds are living dinosaurs," birds directly descended from, or have shared common ancestry with one of the Cretaceous theropods that lived in the Jurassic Period. However, Feduccia, who proposes that "birds are not dinosaurs," states that the dinosaur-bird theory has multiple problems: He says that while dinosaurs are cold-blooded creatures, whose body temperature follows the temperature of their environment, birds are warm-blooded and able to maintain almost constant body temperature.[114] Indeed, because of their high metabolic rate, birds have a body temperature that is even higher than mammals. Also, bird lungs are designed to meet the high metabolic needs of flight, with a set of nine interconnecting flexible air sacs that allow one-way airflow through their lungs. But dinosaurs, like reptiles and mammals, have lungs which breathe in and out—in two directions. The fossil of a true flying bird, named *Protopteryx fengningensis* reportedly dates to 120 million years much before the appearance of dinosaurs.[115] The fossils also revealed that birds were part of dinosaur's diet.[116] "Archaeopteryx is often depicted as a terrestrial predator, with a sickle claw, despite evidence that Archaeopteryx was arboreal."[117] Now Archaeopteryx is considered as "a fully fledged bird, capable of flight, not an intermediate running

---

113 Janvier and Clément, "Muddy tetrapod origins," *Nature*, 40–41.
114 Feduccia, *The Origin and Evolution*, 1999.
115 Zhang and Zhou, "A Primitive," *Science*, 1955–1959.
116 Boardman-Pretty, "First evidence," *New Scientist*, Nov. 2011.
117 Feduccia, *"Birds are dinosaurs,"* Auk 119, 1187–1201.

or glider."[118]

Pakicetus & Ambulocetus

In 1983, paleontologists discovered Pakicetus, meaning "whale of Pakistan," because its bones were found in Pakistan. Some regarded it as the first whale because it had the long skull shaped like a whale. However, few researchers found almost the whole skeleton of *Pakicetus* and it was a 100 % land animal.[119] In 1992, some other paleontologists found a 49-million-year-old fossil with a long skull in a desert in Pakistan and named it as *Ambulocetus natans*, or 'walking whale that swims.'

Based on *Pakicetus* and *Ambulocetus natans*, some biologists believe that whales evolved from certain land mammals around 55 million years ago, although the transformation from a four-legged land mammal to a marine mammal would have taken only 10 million years, a very short span in evolutionary time scale. An interesting article goes this way:

> Ages after some adventurous fish left the sea and planted the flag of vertebrate animal life on land their descendants had it both ways as amphibians and then completed the epic transition, evolving into terrestrial reptiles, mammals and birds. But something about the water must have kept beckoning, until a few irredentists among the mammals did eventually reclaim a place in the sea. Most prominent of these mammals are the whales. Although they may swim the oceans with power and grace, these leviathans are more closely related to the camel and cow than any fish in their wake. Their anatomies retain vestiges of the four-legged land animals in their ancestry, the ones that began the bold return to the sea more than 50 million years ago. But even Darwin much later could not fathom the evolutionary steps by which some land mammals had become whales. It was a big evolutionary plunge, one that has eluded documentation for many decades.[120]

Another reason why some scientists believe that whales evolved

---

118 https://www.allaboutscience.org/archaeopteryx.htm.
119 Thewissen et al., "Skeletons of terrestrial," *Nature*, 277–281.
120 Wilford, "How the Whale Lost," *The NY Times*, May 1994.

from land animals is that whales have pelvic bones, which seem redundant, as they do not need them for locomotion. "Why would a whale, which lacks lower limbs and doesn't need pelvic bones to move, have pelvic bones that are homologous to creatures that do need pelvic bones to move? Similar homologies exist for snakes and legless lizards. Once again, the only explanation that makes sense is if these creatures evolved from a common ancestor along with all the other tetrapods."[121]

Thus the pelvic bones of whales, coccyx at the end of vertebral columns in humans, eyes of blind fishes like cave fishes, the vestigial organs in a species that is not being used are interpreted as evidence for evolution. However, over the years nearly all of the organs once thought to be useless have been found to be functional. As per a study made by Smithsonian, the pelvic bone of whale has an important function in reproduction and sexual selection.[122] Males have a larger pelvic bone that would give more maneuverability while tackling the logistically difficult task of mating in the water. Casey Luskin lists out some of the structures that were previously incorrectly considered to be vestigial, but are now found to have some important purpose.[123] They are—the tonsils that help to fight infection; coccyx (tailbone) which is a vital part of our skeleton, which plays a part in connecting the tendons, ligaments and muscles that support the pelvis; thyroid, which is vital for regulating metabolism; and, the appendix, which is now known to perform significant functions, including producing white blood cells, providing a store of beneficial bacteria, and further significant roles during fetal development. Nowadays paleontologists are very wary of using vestigial organs as an argument favoring evolution.

Neanderthals and similar remains

Neanderthal Man

The bones of "Neanderthal Man" were first found in 1856 in the Neander Valley, in Germany. In Latin, their name is *Homo neanderthalensis* or *Homo sapiens neanderthalensis*. According to the

---

121 Cline, "Whale Pelvis," *ThoughtCo*, Mar. 2017.
122 Thompson, "Promiscuous Whales," *Smithsonian*, Sept. 2014.
123 Luskin, "Problem 10," *Evolution News*, February 2015.

Smithsonian National Museum of history, Neanderthals lived around 400000 to 40000 years ago. When the first specimen Neanderthal 1 to be recognized as an early human fossil was discovered, "scientists had never seen a specimen like it: the oval-shaped skull with a low, receding forehead and distinct brow ridges, the thick, strong bones."[124] Paleoanthropologist Eric Trinkaus says, "They may have had heavier brows or broader noses or stockier builds, but behaviorally, socially and reproductively they were all just people."[125] Also, Neanderthals were believed to be "highly intelligent, able to adapt to a wide variety of ecological zones, and capable of developing highly functional tools to help them do so. They were quite accomplished."[126] Many scientists believe that Neanderthals were very much similar to modern humans. "The numerous associated skeletons of H. neanderthalensis indicate that their body shape was within the range of variation seen in modern humans."[127]

Surprisingly, no one knows where Neanderthal man came from or where he went, as he disappeared about 35,000 to 40,000 years ago. Their demise is an ongoing mystery. Possible causes for their extinction include climate change, losing out in the competition for resources, perhaps due to limited intelligence, or even rival humans killing them off. Some scientists theorized that the volcanic eruption, known as the Campanian Ignimbrite super-eruption, which took place near modern-day Naples in Italy some 40,000 years ago, covered the area in lava and ash and lowered temperatures throughout Europe, causing the final demise of the Neanderthals.[128] Some believe that the Neanderthals were assimilated within the expanding human population because some human-like characteristics have been found in late Neanderthal fossils. Some also believe that, since Neanderthals may have suffered from psoriasis and from other modern medical disorders such as Crohn's disease, the legacy genes would have been inherited by Europeans now, who have roughly two percent

---

124 Smithsonian, "What does it mean?."
125 Lemonick, "Bit of Neanderthal?." *Time Magazine*, April 1999.
126 Alper, "Rethinking Neanderthals," *Smithsonian*, March 2012.
127 Wood, and Collard, "The human genus," *NCBI*, Apr. 1999.
128 Ghose, "Did a Volcano Wipe," *Live Science*, Dec. 2014.

Neanderthal DNA.[129]

Summarizing: the differences between these human-like members of the genus Homo can be explained as micro-evolutionary effects of "size variation, climatic stress, genetic drift, and differential expression of [common] genes . . . The major members of Homo—such as erectus and the Neanderthals (Homo neanderthalensis)—are very similar to modern humans. They're so similar to us that some paleoanthropologists have classified erectus and neanderthalensis as members of our own species, Homo sapiens."[130]

Denisovans

Denisovans were named after the place where the finger bone of a girl and, later, three tiny physical remains—a tooth, a toe bone, and a finger bone—were discovered in a Siberian cave.[131] Some of the speculations were: people living in certain parts of Northern Australia might be sharing 4 to 6 percent of their DNA with Denisovans; since the tooth was far bigger than those of either Homo sapiens or Homo Neanderthals, Denisovans might have been huge; and they seem to have coexisted and even interbred with Neanderthals.

Ardi

Figuring out the story of human origins is like assembling a huge, complicated jigsaw puzzle that has lost most of its pieces. Many will never be found, and those that do turn up are sometimes hard to place. Every so often, though, fossil hunters stumble upon a discovery that fills in a big chunk of the puzzle all at once — and simultaneously reshapes the very picture they thought they were building. The path of just such a discovery began in November 1994 with the unearthing of two pieces of bone from the palm of a hominid hand

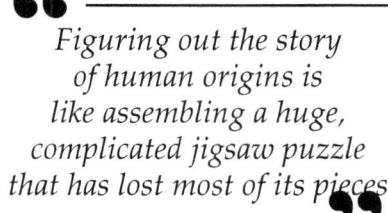

Figuring out the story of human origins is like assembling a huge, complicated jigsaw puzzle that has lost most of its pieces

---

129 Rincon, "Neanderthals gave," *BBC News*, Jan. 2014.
130 Gauger et al., *Science and Human Origins*, 73, 116.
131 Than, "New Type," *National Geographic News*, Mar. 2010.

in the dusty Middle Awash region of Ethiopia. Within weeks, more than 100 additional bone fragments were found during an intensive search-and-reconstruction effort that would go on for the next 15 years and culminate in a key piece of evolutionary evidence revealed: the 4.4 million-year-old skeleton of a likely human ancestor known as *Ardipithecusramidus* (Ardi).[132]

Scientists were excited at the unearthing of the bones of many parts of Ardi because this could provide a clue about our last common ancestor. However, the discovery brought more confusion than consensus. Chimpanzees seem to be our closest living primate relatives, so some would expect to find a transitional form of a human with chimp features. But Ardi had little resemblance to chimps. Ardi was much more human-like than chimp-like.

> The fossil analysis of Ardi is rewriting the story of human origins and changes the notion that humans and chimps, our closest genetic cousins, both trace their lineage to a creature that was more like today's chimp. Rather, the research suggests that our predecessors lacked tusk-like canines to brawl with, or hand-like feet to swing from trees, dashing the popular image of a chimp-like start for homo sapiens but our common ancestor was a walking forest forager more cooperative in nature. This means that while humans didn't diverge much from their evolutionary ancestors, chimps and gorillas look like really special evolutionary outcomes. It says, 'We're going to have to rewrite the textbooks on human origins.'[133]

Boskop man

Boskop man was named after some skull fragments were found in Boskop, a small town from the east coast of South Africa in 1913. It was reported that "the cranial capacity must have been very large . . . The Boskop skull, it would seem, housed a brain perhaps 25 percent or more larger than our own."[134] Might be they had a higher IQ? While some believed that such skulls could have been the result of some aberration caused by some disease like hydrocephalus, others believed it as a new species, due to some distinct features.

---

132 Lemonick and Dorfman, "Ardi Is," *TIME*, Oct. 2009.
133 Than, "Fossils Could Force," *Live Science*, Aug. 2007.
134 Lynch, "What Happened," *Discover magazine*, Dec. 2009.

The face of these people would have been small, childlike, as they had a face to cranium ratio of 1/5th, a European adult has a ratio of 1/3rd. The other features of the face were all smaller too; small teeth, nose, cheeks, and jaw. It appeared that there was a group of humans that lived 12-10,000 years ago in South Africa that had huge brains and tiny faces. The human of the future, living in the past . . . in the 1950's the legitimacy of the finds were called into question. It was said that the paleontologists were finding larger skulls among other human remains and 'cherry picking' them out as Boskops. Claims were that the skulls were within a normal range variance for Homo sapiens, in addition the pieces of skull found were thicker than average and so would make estimating the brain size difficult and inaccurate.[135]

The present understanding of Boskop man is that he is probably a distinctive form of modern man.

Homo floresiensis

About "the "little people" found a few years ago in a cave on the island of Flores in Indonesia, an article says, "The Australian and Indonesian discoverers concluded that one partial skeleton and other bones belonged to a now-extinct separate human species, *Homo floresiensis*, which lived as recently as 18,000 years ago. The apparent diminutive stature and braincase of the species prompted howls of dispute. Critics contended that this was not a distinct species, but just another dwarf-size Homo sapiens, possibly with a brain disorder."[136]

Other Transitional Forms

Roger Lewin cites eight main controversial specimens, like the *Taung child*, a southern ape from Africa, that was once considered as a human ancestor; *the Piltdown man* with a human skull and Orangutan Jaw; the tooth of the *Nebraska man* which actually belonged to a pig, and Ramapithecus, once thought of as human ancestor, but now found to be more related to apes. Paleoanthropologists have uncovered remains from more than 300 individuals, of a species called *Australopithecus*

---

135 Getonthetrain, "Boskop Man," *Steemit*.
136 Wilford, "The Human Family Tree," *The NY Times*, June 2007.

afarensis.¹³⁷ They lived in East Africa between 3.8 and 2.9 million years ago. Lucy belongs to this category. However, it turned out that many of the features of *Australopithecus afarensis* were chimpanzee-like.

Some were quick to claim that these remains as ancestors to modern humans, as transitional forms and therefore as "missing links" in our branch of the evolutionary tree, connecting us back to something like apes. But further investigations have shown that they walked upright, existed alongside other "hominins," have many similarities to modern humans, but do not have many similarities to apes. Hence many scientists believe that interpreting such hominin fossils as historical, physical evidence of our common ancestry with apes is not the right approach.

> The fossil evidence for our evolution from apes is actually quite sketchy. Ancient hominin fossils are rare, and they typically consist of bone fragments or partial disarticulated skeletons obtained from different locations around the world and from different geologic strata. They fall into two basic categories: ape-like fossils, and Homo-like fossils. This discontinuity between fossil types is well-known... If our common ancestry with chimps is true, the transition to fully human must include something like the shift from A. afarensis to H. erectus. And here is where the discontinuity lies. H. erectus is the first fossil species with a nearly modern human anatomy and a constellation of traits not seen in any prior hominin. There simply is no good transitional species to bridge the gap.¹³⁸

"The fossil evidence of human evolutionary history is fragmentary and open to various interpretations."¹³⁹

> Hominin fossils generally fall into one of two groups: ape-like species and human-like species, with a large, unbridged gap between them. Virtually the entire hominin fossil record is marked by fragmented fossils, especially the early hominins, which do not document precursors to humans. Around 3 to 4 million years ago, the australopithecines appear, but they were generally

---

137   Lewin, *Bones of Contention*, 1987.
138   Gauger et al., *Science and Human Origins*, 17, 23.
139   Gee, "Return to the planet," *Nature*, 131–32.

ape-like and also appear in an abrupt manner. When our genus Homo appears, it also does so in an abrupt fashion, without clear evidence of a transition from previous ape-like hominins. Major members of Homo are very similar to modern humans, and their differences amount to small-scale micro-evolutionary changes. The archaeological record shows an "explosion" of human creativity about thirty to forty thousand years ago. Despite the claims of evolutionary paleoanthropologists and the media hype surrounding many hominin fossils, the fragmented hominin fossil record does not document the evolution of humans from ape-like precursors, and the appearance of humans in the fossil record is anything but a gradual Darwinian evolutionary process.[140]

Uncovering the Dawn of the Human Race

"Who do you think you are? A modern human, descended from a long line of Homo sapiens? A distant relative of those great adventure-seekers who marched out of the cradle of humanity, in Africa, 60,000 years ago? Do you believe that human brains have been getting steadily bigger for millions of years, culminating in the extraordinary machine between your ears? Think again, because over the past 15 years, almost every part of our story, every assumption about who our ancestors were and where we came from, has been called into question. The new insights have some unsettling implications for how long we have walked the earth, and even who we really are."[141]

Long time ago, roughly 5.5 to 6.5 million years ago, somewhere in a deep woods of Africa, our early forefathers originated from ape-like ancestors and learnt to walk on two legs and then these humans moved from Africa to Eurasia then Australia and finally the Americas by 12,000 to 15,000 years ago, according to the *Out of Africa* theory. But many scientists say an emphatic no to the *Out of Africa* theory, as they point out baffling information about human origin through many studies that combine archaeology, anthropology, and genealogy.

The Dali skull which was recently found in China was dated back to 260,000 years ago. What is interesting about it is that it shows features of both homo- sapiens and homo erectus. This raises a

---

140 Moreland et al., *Theistic Evolution*, chapter XIV.
141 Barras, "Who are you, " *New Scientist*, August 2017.

lot of questions because the current timeline for humans starts at 200,000 years ago based on the out of Africa origin. For the skull to be in China for over 100,000 years before many people believe humans were in China means that our history is not accurate. The implications of this are that humans may not have evolved only in Africa, which would give more support towards the multiregional origin theory because it would match the fact that humans evolved along a single species from ancient humans to modern humans.[142]

The latest discovery of new fossils from China and Asia is rewriting the story of human evolution, as they reveal that our origin could be from Asia and the spotlight should turn more towards the east.[143] The first Indo-French Pre-historical Mission in the Himalayan foothills is the multidisciplinary research program "Siwaliks."

> This program is dedicated to the discovery of cut marks on mineralized bovid bones collected among vertebrate fossils in a fluviatile formation named "Quranwala zone" in the Chandigarh anticline, near the village Masol, and located just below the Gauss–Matuyama polarity reversal (2.58 Ma). Artefacts (simple choppers, flakes) have been collected in and on the colluviums. This important discovery questions the origins of the hominins which made the marks.[144]

These two findings from China and the Siwalik Himalayan foothills in India are remarkable as they point to the fact that the *Out of Africa* story may be wrong or at least incomplete. From the findings, it is easy to conclude that evolution is not a straight line, and it does not provide a clean transition from one form of a species to the next, and the farther we look back and try to trace our human family, we find that those transitions seem less human.

Inferences

Human evolution involves a lot of assumptions so that even Charles Darwin wrote: "The horrid doubt always arises whether the convictions

---

142 Shiwnarain, "Ancient Skull Found," *Science Trends*, Nov. 2017.
143 Douglas, "Asia's mysterious," *New Scientist*, July 2018.
144 Malassé, "The first Indo-French," *Science Direct*, 281–294.

of man's mind, which has developed from the mind of the lower animals, are of any value or at all trustworthy. Would anyone trust the convictions of a monkey's mind, if there are any convictions in such a mind?"[145] According to Jeffrey L. Walling, "the diversity and disparity amidst the lack of transitional strata has reached epic proportions, and it must now be accepted that the supporting evidence just does not exist after an exhaustive and noble effort. Scientific protocol demands that you have a consensus of the data for it to be considered scientifically valid or a scientific certainty. Where there is no consensus and you inadvertently pick and choose values that agree with your theory or your belief then you have moved into the arena of pseudoscience."[146]

> Human evolution is a highly contentious subject and there is no consensus among scientists

Human evolution is a highly contentious subject and there is no consensus among scientists as to whether the controversial specimens unearthed belong to our ancestors or to other species from which we are not actually descended. Scientists are hoping that the DNA revolution will shed further light in discovering our origin.

UNRAVELING THE TREE OF LIFE

Charles Darwin used the metaphor, the tree of life, for expressing the genealogical history of life on Earth and presenting the idea that all life on earth is related by common descent. Later Ernst Haeckel, who was a great admirer of Darwin, devised a number of such evolutionary, or "phylogenetic" trees to describe his ideas about the relationships between species. He believed that the pattern of similarities and differences in species today fits into a tree-like pattern or nested hierarchy, where branching points correspond to the appearance of new traits. He presented the tree as one of the strongest pieces of evidence for the development of life progressing entirely through undirected natural processes.

---

145 Darwin, Francis ed., *The life and letters*, 285.
146 Walling, *His Story: As a Matter of Fact*, 2017.

Initially such phylogenetic trees were constructed using the morphological characters as the sole source of evidence. But many biologists feel that morphological data is not very reliable in predicting the close evolutionary relationship between species. For example, according to the theory of evolution, chimpanzees and humans share a common ancestor; however, according to the Smithsonian National Museum of natural history, there are tremendous morphological distinctions between apes and mankind due to the differences in cognitive ability (skull size), and the skeletal-muscular design permitting bipedal movement and balance in humans.[147] The average weight of the brain of an adult chimpanzee is 384 g (0.85 lb), but a modern human brain weighs 1,352 g (2.98 lb), almost triple the size. Humans have the largest cerebral cortex of all mammals, relative to the size of their brains. Also, the human brain is very complex, capable of solving problems and creating abstract ideas and images, capable of solving problems and creating abstract ideas and images. It can store many decades' worth of information, collect and process information, then deliver output, in split seconds. Also, humans walk upright, unlike chimps, and they can do a much greater range of things than any apes can do. If one compares humans and apes, their dissimilarities are more, in terms of advanced language skills, spirituality, and emotions like intimacy and human love.

"In examining claims of similarity between animals and humans, one must ask: What are the dissimilarities? This approach prevents confusing similarity with equivalence. We follow this approach in examining eight cognitive cases—teaching, short-term memory, causal reasoning, planning, deception, transitive inference, theory of mind, and language—and find, in all cases, that similarities between animal and human abilities are small, dissimilarities large."[148]

Also, it is possible for two, relatively unrelated species, which have to adapt to living in similar conditions, to end up developing similar features e.g. the sugar glider and the flying squirrel. Bats and whales use echolocation to make high-pitched noises and listen for returning echoes to navigate and hunt, but this trait is absent in their

---

147 Smithsonian, "What does it mean?."
148 Premack, "Human and," *Proc Natl Acad Sci USA*, 13861–13867.

distant common ancestor. To explain such phenomena, convergent evolution has been proposed (as opposed to divergent evolution). But even with convergent evolution, it is quite difficult to explain how and when one family of bats, called the fruit bats, that does not have these echo-locators, lost this capacity along their way.

Building Phylogenetic Trees

In the late 1960s, the availability of data on both protein sequence and DNA sequence allowed scientists to use genetic information to compare one species with another by comparing the sequences of a protein and counting the number of positions where the two sequences differ. This is the molecular clock method, in which the differences in protein sequence between two species are assumed to be proportional to the time elapsed since they diverged. "If you have a start date and know how many genetic mutations there are in a certain period of time, then you can calculate the last time a sample of DNA sequence diverged and the ages of various groups of organisms can thus theoretically be calculated."[149] Since closely related species typically have few sequence differences as compared to less related species, the phylogenetic trees can be built using the comparison of the sequences of genes or proteins among species. Nowadays, a combination of data from a variety of sources, including morphological data, fossil data and molecular data (especially DNA sequences) is used to construct phylogenetic trees. To reconstruct the evolutionary history of species, large amounts of phylogenetic data are available. Due to the increased computational power available today, it is quite easy to analyze and manipulate extremely large phylogenetic data sets and hence research in this field is gaining much interest.

Dating Phylogenetic Trees

Dating a phylogenetic tree involves calibrating the tree and estimating the time of divergence of species. Normally the fossil data and the molecular clock data are used for finding the evolutionary time scale, that is, for estimating of the age of each clade (specific biological group).

Estimation based on fossil data

For estimating the age based on fossil data, biologists face two main

---

149 Simon Ho, "The Molecular Clock," *Nature Education*, 168.

problems: first, the incompleteness of the fossil record; and, second, problems associated with the accuracy of fossil dating techniques. A recent article speaks about three techniques for estimating the ages of fossils:[150]

a. *Biostratigraphy*: This is one of the earliest scientific dating methods. Layers of rock build one atop another—find a fossil or artifact in one layer and you can reasonably assume that it is older than anything above it.

b. *Radiocarbon dating*: Certain unstable isotopes of trace radioactive elements in both organic and inorganic materials decay into stable isotopes. This happens at known rates. By measuring the proportion of different isotopes present, researchers can figure out how old the material is. In radiocarbon dating or carbon-14 dating, carbon-14 in bones or a piece of wood is measured to calculate the date.

c. *Thermoluminescence*: Over time, certain kinds of rocks and organic material, such as coral and teeth, are very good at trapping electrons from sunlight and cosmic rays pummeling Earth. Researchers can measure the amount of these trapped electrons to establish an age. Silicate rocks, like quartz, are particularly good at trapping electrons.

However, the fossil dating techniques face challenges in accurately determining the age of the fossils, such as the risk of contamination in radiocarbon technology and the need for calculating the rate at which electrons were trapped in the thermoluminescence method.

Estimation based on Molecular Clock

Like the fossil dating techniques, the molecular clock method also has major problems:

a. **Incompleteness of the fossil record**—the molecular clock is usually calibrated using fossil data, or geological events that have shaped biogeography and this gives rise to large discrepancies in dating. "The fossil record is well known to be incomplete. Read literally, it provides a distorted view of the history of species divergence and extinction, because different species have different propensities to

---

150   Tarlach, "Everything Worth," *Discover Magazine*, June 2016.

fossilize, the amount of rock fluctuates over geological timescales, as does the nature of the environments that it preserves. Even so, patterns in the fossil evidence allow us to assess the incompleteness of the fossil record."[151]

b. *Varying Estimation by the molecular clock method*—the molecular clock is found to tick "too fast," suggesting an older lineage than the fossil record would suggest. Pulquério and Nichols report: "twofold differences have been reported between the dates estimated from molecular data and those from the fossil record; furthermore, different molecular methods can give dates that differ 20-fold."[152] They have also stated in the same article that, when one of their colleagues submitted an article using the molecular clock method to estimate the date, he was surprised when a referee advised him that his estimate was a factor of ten wrong. How could such a large discrepancy come about? That researcher had used a timescale calculated by making comparisons between different species. On the other hand, the referee used a "pedigree" study, with more closely-related samples. There is a fundamental problem. Using slightly different approaches to the molecular clock method, estimated ages vary by a factor of ten. So it is clear that we do not know the true dates when species diverged.

c. *Disagreement among researchers*—another article says "For the past several years, there have been two main genetic methods to date evolutionary divergences—when our ancestors split from Neanderthals, chimpanzees, and other relatives. The problem was, the results of these methods differed by nearly two-fold. By one estimate, modern humans split from Neanderthals roughly 300,000 years ago. By the other, the split was closer to 600,000 years ago. Likewise, modern humans and chimps may have diverged around 6.5 or 13 million years ago. Puzzled by this wild disagreement, researchers with diverse expertise have been studying it from different angles."[153]

---

151   Donoghue, "The evolution," *Phil. Trans. of the Royal Soc. B*, June 2016.
152   Pulquério, "Dates from molecular," *Trends in Ecology & Evol.*, 180–184.
153   Alex, "Why we're closer," *The Guardian*, Dec 2016.

d. Different genes predict conflicting results—Syvanen carried out research on 2000 genes of six very diverse animals and, theoretically, he should have been able to use the gene sequences to construct an evolutionary tree showing the relationship between the six animals; but he couldn't.[154] The problem was that the different genes told contradicting evolutionary stories. "Evolutionary trees from different genes often have conflicting branching patterns."[155] One gene gives one version of the tree of life, while another gene gives a highly different, and conflicting, version of the tree.

Also based on molecular data, Prokaryotes do not fit a hierarchical scheme. Doolittle says, "Indeed, for prokaryotes, molecular data have falsified the [tree of life] hypothesis."[156] Summing up—"phylogenetic conflict is common, and frequently the norm rather than the exception."[157] Such conflicts have become a norm, giving radically different results from different estimates. Darwin's suggestion of referring to a single progenitor as the common ancestor to a clade has been proved wrong by the modern evolutionary theory. "The last universal common ancestor may have comprised a population of organisms with different genotypes that lived in different places at different times."[158]

Winston Ewert has published a thought provoking paper, in which he proposes that life is best explained, not by Darwin's hypothesis of an ancestry tree, but by a modern design-inspired hypothesis of a dependency graph.[159] He writes that life exhibits an approximate nested hierarchy pattern rather than forming an exact nested hierarchy. Though the resemblance to the nested hierarchy pattern is weak, it must be explained as it is undeniably present. Since the tree of life is not able to explain this nested hierarchy pattern, those who hold on to the tree of life are forced to resort to various mechanisms for explaining the deviations from the hierarchy, such as horizontal

---

154 Syvanen, "Evolutionary," *Annual Review of Genetics*, 341–358.
155 Degnan, "Gene tree," *Trends in Ecology and Evolution*, 332–340.
156 Doolittle, "The practice," *Phil. Trans. of the Royal Society*, 2226.
157 Dávalos et al., "Understanding phylogenetic," *NCBI*, Aug 2012.
158 Theobald, "A formal test," *Nature*, 226.
159 Ewert, "The dependency graph of life," *BIO-Complexity*, 1–27.

gene transfer, incomplete lineage sorting, differential gene loss, gene resurrection, gene flow, and convergent evolution. Ewert says that his new hypothesis that interprets the pattern of similarities in different groups of species as fitting into what are referred to in computer science as dependency graphs is able to explain the approximate nested hierarchy pattern much better.

Regarding Ewert's model, Brian Miller says that, as opposed to the evolutionary tree model that requires, at least for complex life, that every species should link back to a single most recent common ancestor (MRCA) with other species in the same group (clade), the dependency graph model interprets the relationships not in terms of most recent common ancestors but in terms of shared modules.[160] In the dependency graph model, similarities appear to be the result of a designer reusing design modules in different species to meet common goals. This model allows research based on the intelligent design framework. This is because the intelligent design framework appeals to functional constraints or common design instead of common descent for explaining the similarities. "An intelligent cause may reuse or redeploy the same module in different systems, without there necessarily being any material or physical connection between those systems. Even more simply, intelligent causes can generate identical patterns independently."[161]

Tracing our Common Ancestor

Human evolution has always been a contentious subject. In 2005, once the chimpanzee genome sequence was available, scientists could compare it to the human genome and it seemed that 98.5 percent of human DNA is identical to that of chimpanzees. According to this research, it appears only a small portion of our DNA is giving us human traits and it is just a handful of genes that are so powerful in making us humans. "For decades scientists have known that at least 98 % of human DNA is identical to that of chimpanzees. Now they have at last begun to explore which genes separate us from the apes. We humans like to think of ourselves as special, set apart from the rest of the animal kingdom by our ability to talk, write, build complex structures,

---
160   Miller, "BIO-Complexity." *Evolution news,* July 2018.
161   Nelson and Wells, "*Homology in Biology,*" 2003.

and make moral distinctions. But when it comes to genes, humans are so similar to the two species of chimpanzee that physiologist Jared Diamond has called us the third chimpanzee."[162]

> *This idea that humans and chimps are closely related is accepted only by a limited group of scientists*

Richard Dawkins is quoted as saying, "We admit that we are like apes, but we seldom realize that we are apes." However, this idea that humans and chimps are closely related is accepted only by a limited group of scientists, due to the various contradictory findings in two main areas:

a. The proportion of human and chimpanzee Genome that is identical to one another.

b. Chimps may not be the only contender to be our closest relatives—there are others.

Human and Chimpanzee Genome

A recent study contradicts the claim that human DNA is 98.5 percent identical to that of chimpanzees—"For decades, scientists have agreed that human and chimpanzee DNA is 98.5 % identical. A recent study suggests that number may need to be revised. Using a new, more sophisticated method to measure the similarities between human and chimp DNA, the two species may share only 95 % genetic material."[163] Another finding says—"Along the lineage leading to modern humans we infer the gain of 689 genes and the loss of 86 genes since the split from chimpanzees, including changes likely driven by adaptive natural selection. Our results imply that humans and chimpanzees differ by at least 6 % (1,418 of 22,000 genes) in their complement of genes, which stands in stark contrast to the oft-cited 1.5 % difference between orthologous nucleotide sequences."[164]

In humans, each cell normally contains 23 pairs of chromosomes,

---

162 Gibbons, "Which of Our Genes," *Science*, 1432–1434.
163 Pickrell, "Humans, Chimps," *National Geographic News*, 2002.
164 Demuth et al., "The Evolution of Mammalian," *PLOS*, Dec. 2006.

while chimps and great apes have 24 pairs of chromosomes. If humans have 23 pairs and chimps have 24, how did we evolve from a common ancestor? Perhaps, after our branches diverged some six million years ago, our ancestors could have lost one pair of chromosomes. But some biologists argue that, instead of losing, we combined a couple of them. Chromosomal fusion is common, and so the human chromosome 2 should have been derived by the fusion of two primate chromosomes (2a and 2b).[165] However, conclusions from various researches contradict. Regarding telomere length a study predicts—"At the end of each chromosome is a string of repeating DNA sequences called telomeres. Chimpanzees and other apes have about 23,000 base pairs of DNA at their telomeres. Humans are unique among primates with much shorter telomeres only 10,000 long."[166]

> The area of the alleged fusion in chromosome 2 in humans is dissimilar to the telomeres at the ends of chromosomes 2a and 2b in chimpanzees. The alleged fusion would have involved an unprecedented head to head fusion event. In addition, there are active genes in the alleged fusion area. The genes in chimpanzee chromosomes 2a and 2b are dissimilar from the genes found in human chromosome 2. And even if a fusion event had occurred, it could just have been in the human line and would not necessarily be evidence for common descent with chimpanzees. Moreover, telomeric DNA is common in mammalian genomes, so finding it in human chromosome 2 is hardly unique.[167]

Research on junk DNA—the DNA that does not code for a protein, suggest that the similarities between chimp and human genes lie mainly in the regions within DNA (2 percent) that are involved in protein coding, and not in the vast amount of junk DNA (98 percent).

"The things that separate chimpanzees from humans appear obvious on the surface. Humans are more graceful ice skaters, and we wear tuxedos with more panache than our closest primate relatives. We are, however, strikingly similar species on the level of our genes.

---

165 O'Neill, "Homozygosity," *Journal of Human Genetics*, 559–564.
166 Kakuo et al., "Human Is a Unique," Biochem, 308–314.
167 Tomkins, "The chromosome," *Journal of Creation*, 111–117.

The parts of our DNA that contain instructions for making proteins—the building blocks of our bodies—differ by less than 1%, but protein-coding genes are only a small part of our genomes. Some of the biggest differences between humans and chimps lie in the DNA that resides outside of genes."[168]

Also, the presence of this junk DNA, or pseudogenes, or selfish tandem repeat DNA in the human genome has been used by Richard Dawkins and a few others as evidence for unguided evolution.[169] However, few scientists like Richard Sternberg have found extensive evidence of function for repetitive DNA; they found that the non-protein-coding DNA was involved in cell proliferation, cellular stress responses, gene translation, regulating DNA transcription, aiding in folding and maintenance of chromosomes, DNA repair, helping to fight disease and regulating embryological development and in numerous other functions.[170] Around 400 international scientists belonging to the ENCODE project studied the functions of non-coding DNA in humans and stated: "These data enabled us to assign biochemical functions for 80% of the genome, in particular outside of the well-studied protein-coding regions."[171]

Based on all the above studies, one can conclude that the finding that human and chimpanzee are 98.5 percent identical based on DNA, is an over-estimate and needs to be revised. Also, it is becoming increasingly clear that the more the genome is researched, the more functions are discovered for non-coding DNA, which may not be "junk" after all.

"Taken together, these species-specific genetic differences contribute to our anatomical and physiological differences with chimps. In addition, there is not enough evolutionary time for all these coordinated changes to have happened by the mutation/selection process. Thus the evidence for common ancestry put forward by various scientists are not as solid as they might seem. The more we

---

168   Zorich, "Is Junk DNA" *Scientific American*, Jan. 2018.
169   Dawkins, *A Devil's Chaplain*, 2004.
170   Sternberg, "On the Roles," *NY Acad of Sciences*, 154–188.
171   ENCODE, "An integrated encyclopedia," *Nature*, 57–74.

learn about our human genome, the more it seems to be brilliantly and uniquely designed."[172]

Any contenders to Chimps?

Many researchers say that there are many contenders to chimps in becoming the closest ancestor, after assembling the complete genome of chimpanzees, gorilla and orangutan and even bonobos, an African ape. The findings are listed below:

a. Based on the study of a research group—"The chimpanzee genome differs from the bonobo genome by about 0.3 %, which is one-fourth the distance between humans and chimps. Yet chimps and bonobos have radically different social systems, cultures, diets and mating systems."[173] The study also found that a small bit of our DNA about 1.6 percent, is shared with only the bonobo, but not chimpanzees. Now the chimpanzee has to share the distinction of being our closest living relative in the animal kingdom with the bonobo.

b. These studies have "prompted researchers to speculate whether the ancestor of humans, chimpanzees, and bonobos looked and acted more like a bonobo, a chimpanzee, or something else—and how all three species have evolved differently since the ancestor of humans split with the common ancestor of bonobos and chimps between 4 million and 7 million years ago in Africa."[174]

c. A team of researchers have completed the genome sequence for the gorilla, the last genus of the living great apes. "The team show that much of the human genome more closely resembles the gorilla than it does the chimpanzee genome . . . 15 % of the human genome is closer to the gorilla genome than it is to chimpanzee, and 15 % of the chimpanzee genome is closer to the gorilla than human . . . gorillas share many parallel genetic changes with humans including the evolution of our hearing. Scientists had suggested that the rapid evolution of human hearing genes was linked to the evolution of language."[175]

---

172   Nelson, "Beautiful Monster" Evolution news, Nov. 2017
173   Prüfer et al, "The bonobo genome," *Nature*, 527–531.
174   Gibbons, "Bonobos Join Chimps," Science mag., June 2012.
175   Science News, "What have we got," March 2012.

d. A group of researchers argue that this orangutan-like ancestor of humans is native to Southeast Asia and not to Africa and they may be our closest relatives, and not Chimpanzees because the DNA evidence cited by many scientists was based on only a small percentage of the human and chimp genomes. "Humans share at least 28 unique physical characteristics with orangutans but only 2 with chimps and 7 with gorillas . . . features shared by both orangutans and humans include thickly enameled molar teeth with flat surfaces, greater asymmetries between the left and right side of the brain, an increased cartilage-to-bone ratio in the forearm, and similarly shaped shoulder blades. A hole in the roof of the mouth that was supposedly unique to humans is also present in orangutans . . . The team also highlighted Orangutan-type traits in the teeth and jaw remains of ancient fossil apes from Africa and Europe."[176]

In the light of this, human origin needs a radical re-think. The argument for our common ancestry with ape-like creatures based on the similarities between two complex structures is not the right argument, as the amount of genetic change that would be needed to actually accomplish the proposed evolutionary transitions in the given amount of time by any unguided process is impossible—

> Two questions then arise: (1) How many mutations would it take to turn anaustralopithecine species into a Homo erectus? And (2) If there are only one and a half million years between A. afarensis and H. erectus, can neo-Darwinism produce the necessary changes in the time allotted? . . . But six million years is the entire time allotted for the transition from our last common ancestor with chimps to us according to the standard evolutionary timescale. Two hundred and sixteen million years takes us back to the Triassic, when the very first mammals appeared. One or two mutations simply aren't sufficient to produce the necessary changes—sixteen anatomical features—in the time available. Many of the anatomical changes seen in H. erectus had to occur together to be of benefit. Individually they would be useless or even harmful. So even if a random mutation or two resulted in one change, they would be unlikely to be preserved. And getting

---

176   Owen, "Orangutans May," *National Geographic News*, June 2009.

all sixteen to appear and then become fixed within six million years, let alone the one and a half million that it apparently took, can't have happened through an unguided process.[177]

Given this timeframe, it is extremely improbable and impossible, for us to have evolved from hominin ancestors by a gradual, unguided process. If we have not evolved from hominin ancestors, then how did we come into existence? Population genetics is used in academia to infer that we share a common ancestry with apes. But scientists from the Discovery Institute have proposed an Alternative Population Genetics Model—a unique origin model, which suggests that the humanity arose from one single couple and that there, is a real possibility of a founding first pair.[178]

> *it is extremely improbable and impossible, for us to have evolved from hominin ancestors by a gradual, unguided process*

Inferences

The theory of common ancestry is not well substantiated, because it has to extrapolate to fill such enormous gaps. The evidence presented here suggests that there are huge, unbridgeable gaps in the scientific understanding of evolution and how life started from non-life. The "Tree of Life" is a powerful image, familiar to many from their school textbooks, with a single-celled ancestor forming the base of the trunk and with humans as one among many modern species at the tips of the branches. But, after examining evidence, one can conclude—"for a long time the holy grail was to build a tree of life. But today the project lies in tatters, torn to pieces by an onslaught of negative evidence."[179]

CONCLUSION

Many are aware that Darwinian evolution is a controversial topic because of religious beliefs, but few realize that it is also controversial because of disagreement based on scientific evidence. Almost all

---

177   Gauger et al., *Science and Human Origins*, 24, 25.
178   Moreland J.P. et al., *Theistic Evolution*, chapter XVII.
179   Lawton, "Why Darwin was wrong," *New Scientist*, Jan. 2009.

scientists believe in small-scale changes in a population of organisms, which is called micro-evolution. However, there are many scientific skeptics when it comes to universal common descent, the origin of life, and the unguided process of evolution—that is, macro-evolution, involving natural selection, along with random mutation, giving rise to large-scale changes and the complexities of various life forms. In this chapter, scientific objections to four significant aspects of neo-Darwinism have been presented namely:

a. The origin of life on earth (Abiogenesis)

b. The origins of life's complexity

c. The transitional life forms

d. The reconstruction of the tree of life.

The origin of life seems to be a profound mystery. All the scientific models put forth as explanations are highly speculative and hardly credible and it is very difficult to explain the rise of life spontaneously by chance. The widely-assumed view that the evolution of life has arisen from an unguided process of random mutation and natural selection has been shown to be highly questionable. The fossil record lacks plausible candidates for transitional forms but, rather, shows abrupt explosions of fully-formed new life forms. The idea of universal descent—that all organisms are related to one another because they are all descended from a single, common ancestor—is on careful inspection, far from convincing.

Human evolution has been portrayed in diagrams of branching trees, with progress happening in steady steps. But the story of human evolution seems like a fairy tale, painting an artificially elegant portrait of life's pilgrimage.

> "Man is not merely an evolution but rather a revolution."
> G K Chesterton

PART 3
PROBING THE BACKBONE OF SCIENCE

CHAPTER 4

A SEARCH FOR TRUTH: ACROSS FIELDS OF STUDY

"Science without religion is lame, religion without science is blind."
Albert Einstein

FROM SCIENCE TO PHILOSOPHY

This chapter explores how science does not stand entirely alone as an approach to studying the big questions of life, like our origin and destiny, but depends on the support of philosophy and theology. Science has provided many answers and helped us to understand the world around us better and better—and the pace of scientific development seems to be getting faster and faster. Yet we still face significant questions of what lies, as yet unseen, at the edges of the universe, or hidden within the atom, of our origins in the past and of our destiny in the future. So do we simply need to keep sharpening our scientific tools, building better telescopes and particle accelerators, or is more needed?

Since Einstein's breakthrough, science has advanced at a faster pace. Einstein was on the illusory pursuit of developing a unified field theory that describes all the four forces of nature and demystifies the quantum world for almost 30 long years. Tim Folger recalls an incident in the life of Einstein, when he wrote to a self-taught physicist Muffat in this way, "Dear Mr. Moffat, Our situation is the following. We are standing in front of a closed box which we cannot open, and we try hard to discover about what is and is not in it."[180] Our universe is that closed box—Einstein strove hard to pry off the lid. It looked like Einstein had squandered his genius and wasted his time by chasing vainly after a unified theory that could answer the ultimate question of origin.

> *Einstein had squandered his genius and wasted his time by chasing vainly after a unified theory*

Though he failed in his lonely quest for the Ultimate Theory, it later engaged thousands of physicists around the world, including some of the greatest minds such as Werner Heisenberg and Wolfgang Pauli. Even Stephen Hawking was not immune to its beckoning call, because he was pursuing that one question: What happened at the cosmological singularity when space, time and matter came into existence? Such singularities, such as the Big Bang singularity and black holes, are zones where Einstein's Relativity goes blind, and Quantum Mechanics fails in its attempt to take over. There is no mathematical model to explain it. Hawking believed that if these two theories could be reconciled, it could solve questions such as: What is inside a black hole? What happened at the beginning of the universe? Why is the universe accelerating? However, Hawking died on March 14th on the anniversary of Albert Einstein's birth in 1879, without finding this "theory of everything." Many researchers have spent a great deal of their time and energy pursuing this captivating and seductive scientific prospect. A purely naturalistic explanation for the origin of the physical universe and life and the fine-tuning of the universe seems to simply fail, as most of the theories cannot be tested, verified or falsified.

Why do these theories keep stumbling and why does a theory of everything seem far from reality? Maybe we just don't know enough yet about key phenomena which remain invisible and, while trying to understand the big picture, we might be missing the basic truths. Carlo Rovelli, a theoretical physicist said regarding the theory of everything, "Einstein defined what later became a fundamental problem in physics; but he was missing an ingredient."[181]

---

180   Folger, "Einstein's Grand Quest," *Discover Magazine*, Sept. 2004.
181   Folger, "Einstein's Grand Quest."

A big question pops up in our minds—is there a missing ingredient? Do these theories warrant the postulation of a supernatural ultimate entity? What lies behind the physical reality of our universe? However, since science is limited to the study of the natural world, it could not explain ultimate reality if that reality were a supernatural entity or a metaphysical being beyond our physical reality (the universe and all that is in it). So, how do we find the answer to the question? Is there any other explanation?

PHILOSOPHY OF SCIENCE

Is there an ultimate reality outside the physical reality? Does supernatural reality exist? Where do our moral responsibilities come from? Are there any limits to our knowledge? While I was pondering these questions, I understood that I had to turn to philosophy, which is the discipline that encourages us to think clearly and sharply about ideas and sets of evidence and, in the process, helps us to discover truth by finding rigorous answers to the fundamental questions that arise in everyday life. By delving deep into the philosophy of science, where the scientific enterprise first started, one can discern the hidden assumptions, understand limitations, and construct logical arguments while trying to map the complete picture of the origin of the universe and life.

Scientific enterprise depends on certain presuppositions or assumptions for its functioning. These are philosophical rather than scientific, and hence they can only be defended with an appeal to philosophy. Scientific knowledge is not eternally and infallibly true, but it is limited by uncertainties of various forms and degrees, due to its basic assumptions and limitations. These few basic presuppositions, on which science operates and scientific knowledge relies, are worth acknowledging:[182]

- a. **The world is real**. It is not illusory and, whether we perceive it or not, the physical universe exists.
- b. **The world is rational and ordered**. Humans can perceive and learn correctly how the physical universe works or operates. Also, the

---

182   Nickels, "Nature of Modern," *Illinois State University*, 1998.

person who is investigating should possess rationality on par with the rationality of the world. For example, using mathematics, which is a human invention, we can understand the movement of the stars.

c. *There is a cause behind every finite event.* There are natural causes for things that happen in the world around us. Science assumes that we can learn about gravity by studying the evidence of an apple falling to the ground. But, in contrast, an infinite being need not have a cause, as it is an uncaused cause.

d. *The principle of uniformity*—the working of nature is uniform in both space and time. Under the same conditions, the same cause always produces the same effect in all places and at all times.

Science depends on the human sensory experience of the natural world and many times, this sensory experience is based on our biased perceptions of the world and also, our previous experience, and this is certainly one of the major limitations of science. The other major limitation is that scientific knowledge keeps changing with respect to time, based on new evidence or new ways of thinking. Scientific knowledge many times increases with a radical change of thought. A paradigm shift in the scientific thought process occurred when the scientific methods and practices followed no longer appear adequate to address new phenomena. With time, there is a great chance that theories like inflation can either be integrated into the central framework or can be abandoned and replaced by a better theory. Due to this intrinsic uncertainty or built-in uncertainty in finding the truth, scientific knowledge cannot be considered as absolute. It is necessarily contingent knowledge.

For example, a majority of researchers accept that the cosmological constant, the agent behind the acceleration of the universe, or the vacuum energy density of the universe, must be near zero, but slightly positive, causing the universe to accelerate. Using the data from the Planck satellite, this value is found by fitting the cosmic microwave background's power spectrum with the parameters that determine the expansion of the universe. But when physicists attempted to use basic quantum physics to estimate the cosmological constant the answer came out 120 orders of magnitude higher than the result given by astronomical observations. This drastic error between the experimental value of the cosmological constant and the theoretical prediction shows that research is complex and even the most experienced team of scientists finds it hard to account for all the emerging evidence and can head off in the wrong direction.

SUBJECTIVITY IN SCIENCE

Michael Polanyi made a profound contribution to the philosophy of science by overturning the last few centuries of thought that science was purely objective. He argued that the discovery of scientific reality is a creative act and so, it is guided by passions, strong personal feelings, hunches, imaginings, informed guesses and commitments ("tacit" forms of knowing). Such "tacit knowledge" makes an essential contribution to any kind of scientific knowledge. He claims that "all knowledge is either tacit or rooted in tacit knowledge. Wholly explicit knowledge is unthinkable."[183] "The scientist's decision depends on the strength of the beliefs in the light of which he interprets his observations, and we approve of this decision if we share these beliefs."[184]

For example, most of the elegant equations of Einstein were framed in his mind and what he first produced through his imagination was only later verified with scientific evidence. Subjectivity is part and parcel of the Copenhagen Interpretation of quantum mechanics. Any scientific pursuit is a combination of subjective and objective elements. This twin thread of objectivity plus subjectivity can be seen in almost every discipline of science. Objective refers to objects and events in the world that anyone can, in principle, observe, like the heat of the sun, which is completely unbiased and verifiable. Subjective refers to feelings and experiences that depend on the individual's own particular viewpoint and traits, although it generally also has a basis in reality; it reflects the perspective through which the scientist views reality. The history of science demonstrates clearly that science has been depending on subjective experience and inspiration at the same time as depending on objective experiments and theories. The subjectivity

---

183   Polanyi, *The Tacit Dimension*, 7
184   Polanyi, "Scientific Beliefs," *Ethics*, 30.

and objectivity of science are strongly associated with the nature and image of science itself. It is, arguably, impossible to disassociate human subjectivity from science.

Though subjectivity has a legitimate place in Science, it can lead to the danger of usurping scientific theories to validate one's ingrained worldview. For example, an atheistic worldview seems to underlie Hawking's scientific interpretations. He wrote regarding the theory of everything that, "if we discover a complete theory . . . then we shall all . . . be able to take part in the discussion of the question of why it is that we and the universe exist. If we find the answer to that, it would be the ultimate triumph of human reason—for then we should know the mind of God."[185] The extraordinary fine-tuning of the universe, required for it to be able eventually to support human life, he attributed to chance, but could this be because his mind was not able to conceive of an intelligent design? He writes, "Spontaneous Creation is the reason that there is something rather than nothing, why the universe exists and why we exist. It is not necessary to invoke God to light the blue touch paper and set the universe going."[186] Though he did not believe in God, he was concerned about what will happen to humanity in future and discussed the risks associated with aliens who can destroy us, or artificial intelligence that could lie in wait, or even the Higgs Boson that could become unstable and wipe out the whole universe. Although Hawking did not, during his life, find the elusive Theory of Everything, and although he did not have confidence that there would be an afterlife, I cannot help wondering whether, having passed on, beyond the event horizon, he might now have been surprised, having found the answer he was looking for.

> *Subjectivity in Science can lead to the danger of usurping scientific theories to validate one's ingrained worldview*

---

185   Hawking, *A brief History of Time*, 193.
186   Hawking and Mlodinow, *The Grand Design*, 25.

MOVING ONTO THEOLOGY

By probing the backbone of science, that is, philosophy, I concluded that discussions about the question of the origin of universe and life could not be completed without a strong theological knowledge. There are many great thinkers in history, such as Augustine, Anselm of Canterbury, Thomas Aquinas, William of Ockham, C. S. Lewis, and G.K. Chesterton, who have blended philosophy and theology in their framework of understanding reality. Science and theology have to go together in providing the answer to this profound question about the nature of reality.

In the middle Ages, the lines between the three domains, science, philosophy, and theology, were pretty blurred. "In the Middle Ages, theology and philosophy were a dynamic duo of hub disciplines around which the other sciences were organized. These two areas influenced thinking across the various fields of scholarship."[187] Though the relationship between these three disciplines varied with time, it can be shown that the epistemological (the study of knowledge) foundation on which science stands is strongly anchored in a biblical worldview. This is because Christian theology provided the philosophical underpinnings that science needs. "Historians of science, even non-Christians, have pointed out that modern science first flourished under a Christian world view while it was stillborn in other cultures such as ancient Greece, China, and Arabia."[188] Vishal Mangalwadi writes that "the Bible became the ladder on which the West climbed the heights of educational, technical, economic, political, and scientific excellence."[189] Physicist Paul Davies points out:

> *Science and theology have to go together in providing the answer to this profound question about the nature of reality*

All the early scientists, like Newton, were religious in one way or

---

187   Leydesdorff, "Can scientific," *Jour. of the American Society*, 601–613.
188   Stark, R., *For the Glory of God*, 2003.
189   Mangalwadi, *Truth and Transformation*, 119.

another . . . In the ensuing three hundred years, the theological dimension of science has faded. People take it for granted that the physical world is both ordered and intelligible. The underlying order in nature—the laws of physics—are simply accepted as brute facts. Nobody asks where they come from; at least they do not do so in polite company. However, even the most atheistic scientist accepts as an act of faith that the universe is not absurd, that there is a rational basis to physical existence manifested as a law like order in nature that is at least in part comprehensible to us. So science can proceed only if the scientist adopts an essentially theological worldview.[190]

The culture of the Christian West became the driving force behind the great scientific and engineering feats by affirming all the basic presuppositions needed for science—

a. *The world is real.* The biblical worldview affirms that the natural realm that is felt with our sensory experiences is real, and so is the supernatural realm, which may or may not be felt with sensory experiences. This assumption is rooted in the very character and activity of God in creation. Colossians 1:16 states that "all things were created, both in the heavens and on earth, visible and invisible . . . all things have been created by Him." Hinduism and many forms of pantheism teach that physical reality is an illusion. Christianity, on the other hand, de-deifies nature and proclaims that God is not part of nature, but the Creator of nature. Whereas the gods of nature had been capricious or competing, leading to a hazardous and unpredictable world, the advent of monotheism freed humanity to investigate the world without fear.

b. *The world is rational and ordered.* Humans have the capacity to understand the physical universe because they have been created in the image of the Creator (Genesis 1:26–28). Human logic and reason can be trusted. Since the Creator (the Logos) has stamped His logic upon His creation (John 1:1–3), creation behaves rationally. Francis Schaeffer writes, "Since the world had been created by a reasonable God, [scientists] were not surprised to find a correlation

---

190   Davies, "Physics and mind of God," *First Things*, August 1995.

between themselves as observers, and the thing observed—that is, between subject and object. . . . Without this foundation, modern Western science would not have been born."[191]

c. **Behind every finite event, there is a cause.** "The first verse of the Bible lays the ground rule for causality. "In the beginning, God [the Infinite First Cause] created the heavens and the earth [the finite effect]." This distinction between cause and effect is systematically maintained in the Bible. Even within the Uncaused Cause—the Triune God—there is a real distinction between the Father and the Son, which is the theological and philosophical basis for all true distinctions in the created order. If the distinction between cause and effect is non-existent, all scientific enterprise will cease!."[192]

d. **The Principle of Uniformity**—the Christian worldview affirms the uniformity and regularities of the natural world. As well as day and night, on day 4, God created seasons, with a regular pattern of summer and winter (Genesis 1:14), providing ongoing, regular harvests. Such regularity in nature is the reason for the possibility of the fruitful scientific enterprise.

CONCLUSION

It was in the fertile soil of a biblical worldview that modern science grew and developed, and science still depends on presuppositions which are basically philosophical. Francis Bacon said that—*little philosophy inclineth man's mind to atheism, but depth in philosophy bringeth men's minds about to religion*. While scientific knowledge gives us a reliable understanding of the universe around us and expands the meaning of nature, biblical theology takes this understanding further and provides us with the meaning of our existence. Arthur L. Schawlow, Professor of Physics at Stanford University and winner of a Nobel Prize for Physics, wrote, "It seems to me that when confronted with the marvels of life and the universe, one must ask why and not just how. . . . I find a need for God in the universe and in my own life."[193] So the question, "How did

---

191   Schaeffer, *How Should We Then Live*, 134.
192   Jeyachandran, "Science: God's friend," 5–6,
193   Schawlow, as cited in Margenau and Varghese, 105–106.

this universe come into existence?" also points to this other question, "Why did it happen?" This cannot be answered in an intellectually satisfying manner by science alone. Science cannot be extrapolated to tell us our place in this universe. We need a worldview that gives life meaning, and the biblical worldview provides it beautifully. I strongly believe religion and science complement one another, and should not be pitted one against the other.

Science and theology together help us in understanding the world in a coherent way and thus aid in the emergence of a grand, unified theory of everything.

CHAPTER 5

A SYNTHESIS: SCIENCE AND BIBLICAL THEOLOGY

We have to stop and be humble enough to understand that there is something called mystery.

Paulo Coelho

IN SEARCH OF CONSONANCE

The philosophy of science identifies the undergirding factors behind the scientific enterprise in terms which connect intimately with the biblical worldview. If the God, who is the author of the Bible is also the author of nature, He should have left His signature in both the books (nature and the Bible). I moved on to check whether there are any commonalities and I found overwhelming consonance (parallels) between them, especially concerning mysteries and paradoxes.

UNDERLYING MYSTERIES: SCIENCE

Webster's dictionary defines mystery as something that "resists or defies explanation." Our universe seems to be mysterious. When I started to delve into the incredible world of quantum mechanics, I became fascinated by the range of subjects I encountered which dealt with a host of mysterious phenomena—like wave particle duality, observer paradox, quantum entanglement, the ongoing quest for quantum gravity, etc. As I tried to get to the root of this mystery for understanding the universe's deepest secrets, I encountered a sense of awe and wonder. Along with the amazement, a question also arises about whether we will be ever able to unlock some of the profound mysteries of our universe.

Only 1 part in 20 of the observable universe has been observed experimentally; 4 percent of the mass density in the universe is ordinary matter and the remainder is in the form of dark energy and dark matter, and no one knows what those are comprised of. That is why "Science cannot solve the ultimate mystery of nature. And that is because, in the last analysis, we ourselves are part of nature and therefore part of the mystery that we are trying to solve."[194] Metaphysics is unavoidable in many aspects of quantum mechanics since this theory exceeds the limits of sensible experience. With the Copenhagen interpretation of quantum mechanics, that has the uncertainty principle at its very heart, reality looks incomplete, with many overlapping possibilities. Hence, a few scientists like Louis de Broglie and Bohm suggested that there should be extra stuff besides the wave function, called the hidden variables. However, Bell went on to prove that this is wrong, through his famous Bell's theorem. Most of the unsolved mysteries of science seem to remain unsolved forever.

Extra Dimensional Reality

Mathematics is often called the language of physical sciences, as physical laws can be precisely and beautifully stated in terms of mathematical eqations. Scientists, who probe deep into subatomic structures or into the very early stages of the universe, find elegant mathematical order, although this may require extra dimensions. When we think of a dimension, we think of the three dimensions of space—length, width and depth of all objects in our universe—and the one dimension of time. Beyond these visible dimensions, when scientists say that there may be many more extra dimensions, that seems like something cooked up by science fiction writers.

The idea of invoking extra dimensions started with Theodor Kaluza and Otto Kleinwho showed that, by extending space-time to five dimensions, the then two known fundamental forces, electromagnetism, and gravity, could be unified.[195] But if a fifth

---

194 Planck and Murphy (trans.), *Where is Science*, 217.
195 Kaluza, "On the Unification," Math.Phys..

dimension exists, why is it unobservable? Klein explained that the size of the dimension is very compact, close to the Planck length, around 10^{-30} cm in circumference, and so it is unobservable, though it exists.[196] But ultimately, Kaluza-Klein theory was discarded as they didn't know about the weak and strong forces back then and so their perspective was incomplete. Their idea re-emerged later as string theory, that posits that our universe has eleven dimensions, ten spatial and one temporal. Time is considered to be the eleventh and highest dimension. Though scientists are excited about gleaning any hints of extra dimensions, still now these remain unobserved. They still haven't found that one make-or-break experiment that can furnish some tangible evidence in settling this mystery once and for all.

Unseen, Intangible Reality

Templeton prize winner Bernard d'Éspagnat argues that, since there is a reality independent of our observations, objective reality is forever veiled from human knowledge and science offers us only a glimpse behind that veil.[197] The limits of human observation mean that we are unlikely ever to know the whole universe in detail and we can never know the laws and constraints that govern the occurrence of any event, much less the structure of the whole universe. The reality of quarks, the elementary particles that makeup protons and neutrons in a nucleus, has been established by particle physicists, based on many studies, such as deep inelastic scattering, even though no single quark has ever been detected in isolation in the laboratory. This unseen reality of quarks is due to a property called confinement that binds them so tightly together within the particles that no powerful impact can dislodge them individually.

> Quarks are, in some sense, unseen realities. Nobody has ever isolated a single quark in the lab. So we believe in them, not because we've—even with sophisticated instruments, so to speak—see them, but because assuming that they're there makes sense of great swaths of physical experience. . . . this unseen reality of quarks became an absolutely fundamental aspect of our understanding of the structure of matter. And that remains the

---

196 Klein, "The atomicity of electricity," *Nature*, 516.
197 Gefter, "Concept of hypercosmic," *New Scientist*, 2009.

case. And I, in common with all particle physicists, believe very fervently, in a way, in the reality of quarks. But it's an unseen reality. It's the fact that they give intelligibility to the world that makes us believe that they're actually there.[198]

UNDERLYING MYSTERIES: BIBLE

The Bible presents a superhuman Creator, who is beyond human perception and comprehension. Although the universe and everything in it is confined to a single dimension of time, the Bible speaks of a God who is beyond that dimension of time. He existed from everlasting to everlasting, even before space and time came into being and He will continue to exist, even if this space and time come to an end. The question of who created God is a logical fallacy since He is not confined to our dimensions of time and space. He is the first cause, and there is no necessity for Him to be created. The biblical God is a God who is beyond the boundaries of our perception. He is the unseen reality. Hebrews 11:3 says that "By faith, we understand that the universe was created by the word of God so that what is seen was not made out of things that are visible." Paul says, "We look not to the things that are seen but to the things that are unseen. For the things that are seen are transient, but the things that are unseen are eternal." (2 Corinthians 4:18). Gregory of Nyssa, of the 4th century A.D., explains: "*Aktistos* (Uncreated) God willed and His will materialized into a tangible and a non-tangible *Ktisis* (Creation)."[199] The uncreated God of invisible greatness created both tangible and intangible creation of enormous extent, expanse, volume, variety, and power.

The Bible indicates that the divine intention of this eternal being was to create rational, moral creatures which can be in a conscious relationship with the divine. That lies behind creation. That is also why the unknowable God came down to our planet to reveal Himself. Though God is an infinite, transcendent being, above all goodness, out of mercy and grace, God has chosen to reveal Himself through His Son, in human form, and through Scripture, using human language. It is almost impossible for finite human beings to grasp and ascend to

---

198 Polkinghorne. *Science and Creation*, 113.
199 Elias. "An Orthodox View." Theologia, 607–619.

the heights of the infinite God and it is difficult to find words in the human language to describe Him. This has been expressed in Latin: *finitum non capax infinitum*. God has chosen to reveal Himself, but if He fully revealed Himself, we would not be able to take it in. He is highly exalted, holy, omnipotent and incomprehensible. His name is wonderful, and the journey of knowing this unknowable yet known God is a beautiful journey that one can take while in this world, as it is filled with wonder, awe, and praise.

God asked Job, "Can you discover the depths of God? Can you discover the limits of the Almighty?" (Job 11:7). Only a small bit of God's immensity is revealed to us since man's wisdom is limited. The majestic God is a mystery. Throughout history, one can see the implicit struggle of God's people to find a balance between what is knowable about God and what is unknowable. The Bible does not treat God as a "god of the gaps"—who is pushed into the places that human reason or science cannot currently explain. He is the Creator of the whole world, who ordains and sustains every detail of nature.

"The truth which God has revealed concerning himself in nature and Scripture far surpasses human conception and comprehension. In that sense, dogmatism is concerned with nothing but mystery, for it does not deal with finite creatures, but from beginning to end raises itself above every creature to the Eternal and Endless One himself."[200] Only God, in His infinite wisdom, can keep all reality in focus, whereas human beings are drastically limited in time, space and mental ability. In the Bible, a "mystery" is almost always something that is not immediately clear, but will be made known; it may be something that involves so much information that the human mind finds it impossible to grasp. According to the most common biblical usage, "mystery" denotes the marvelous plan of God that has been revealed. When Daniel was asked to interpret Nebuchadnezzar's dream (Daniel 2:18), the mystery ended up being "revealed" to Daniel (Daniel 2:19, 30, 47). When Jesus speaks about "the mystery of the kingdom of God" (Mark 4:11), although that kingdom remains a secret to others, for the apostles, it is no longer a secret. But still it is a mystery.

---

200 Bavinck, *The Doctrine of God*, 13

The whole fascination of a detective story lies in trying to solve the puzzle, and when one knows the solution the mystery is dissolved—it is no longer a mystery; it has lost its mainspring. But the fascination of many of the New Testament mysteries lies in their peculiar character in that they excite wonder, awe, amazement, astonishment even after they have been revealed. Think again about the mysteries that pertain to the gospel. We understand the good news, and yet it continues to overwhelm us by its elaborate intricacy, its unanticipated beauty, and its stunningly benevolent glory. This is the way a revelation mystery works: we know, and yet the mystery remains. Extensive mystery refers to a quantitative inexhaustibility, a magnitude or an internal complexity that puts some proposed objects of knowledge out of reach. The marvelous elaborateness or beauty of the gospel seems to signal a kind of extensive mystery, where we have an excess of information—but it is too much for us to grasp.[201]

A Christian approach to God's mystery should not involve using it as an excuse to avoid careful theological thought. On the other hand, a mystery should not be treated as simply a puzzle to be solved. Rather, a mystery provides an impetus for seeking answers while also invoking awe, reverence, and worship. At first, when we hear that God is mysterious, we think He must be unknowable. But it is just the opposite. Divine mystery provides the only real hope that we can know God. It is precisely his greatness that makes him accessible to us. God is indeed way beyond our imagination, but, in his love, he has stooped to conquer. Indeed, unless mystery pervades all of life, we are left with a gray, purposeless existence.

> *A mystery is not a contradiction or absurdity. It is beyond reason, but not against reason*

A mystery is not a contradiction or absurdity. It is beyond human comprehension, but not against human understanding. It is beyond reason, but not against reason.

---

201 Boyer and Hall, *The Mystery of God*, 6–7.

PARADOXICAL TRUTHS

Light presents us with a paradox, and this was proved due to two different experiments. While Thomas Young's two-slit experiment demonstrated that light is a wave, Albert Einstein's photoelectric effect demonstrated that it is composed of light particles called photons. Neither of these experimental results can be dismissed and thus came the understanding of wave-particle duality, that light is both a particle and a wave. Light propagates through space as a continuous wave, but somehow, on the way, it exchanges its energy in the form of discrete particles. Later, in 1924, Louis de Broglie came forward and proposed another radical new idea that wave-particle duality applies not only to light but also to matter. The link between the particle properties and the wave properties is Planck's constant h. Wave-particle duality has since then become the cornerstone of quantum mechanics, though physicists have still not fully understood how a single entity such as an electron or a photon can behave both as a particle and as a wave, and that remains a great mystery.

In quantum theory, both the hidden variable theory and the many world theory have dualist ontology—that the physical world contains both the wave function and the additional variables. Modern science has shown that contradictory explanations of reality can apply at the same time, like wave-particle duality. They are not contradictions but paradoxes (that is, two statements that appear to be contradictory initially, but upon further investigation, may both prove to be right).

The Christian worldview also requires us to think in terms of paradoxes. The Bible has many such paradoxes—like the paradoxes of science. Some biblical paradoxes are:

a. **The Trinity**: How can God be both one and also three Persons (Father, Son, and the Holy Spirit)?

b. **The Incarnation**: How can Jesus be, at the same time, both fully God and fully human?

c. **Divine Sovereignty and Man's free will**: How can a man have the freedom to choose and yet be under God's sovereign control? While the Bible says that God is Sovereign and He predestines or chooses people, it also says a man is equally responsible for his actions and that he has to use his free will for the greater good.

d. **God is both Transcendent and Immanent**: While God is transcendent, far beyond our senses; He is also immanent and personal, very close to us. Jeremiah 17:12 says that the Lord, from the very beginning, has been seated on His glorious throne on high, whereas Deuteronomy 31:6 says that the LORD your God goes with you; He will never leave you nor forsake you.

e. **The Crucifixion and Resurrection**: If God is eternal and immortal, how could a divine Being die?

f. **Omnipresence**: How can God be in every place and at all times, past, present, and future?

Though these statements seem contradictory, none of them are contradictions. Scripture says that God's thoughts are not our thoughts and His ways are not our ways (Isaiah 55:3). This indirectly means that His logic, His arithmetic, His truth are not ours. He is beyond reason, not against reason. The biblical paradoxes need not be resolved, because they are beyond human understanding, but they should be embraced, like the paradoxes of modern science. These paradoxes will bring us to a profound recognition of our human limitations and will lead us into a richer awareness of God's majesty and power.

CONCLUSION

Science alone cannot reveal to us the secrets behind the origin of cosmos and life without the support of philosophy and theology. Stephen Hawking stated that "... philosophy is dead," since philosophers have not taken science seriously enough, nor have they kept up with its modern developments.[202] He thought that the fundamental questions about the nature of reality could be resolved with the inputs obtained from the particle physics community and space research. However, Hawking was criticized by many for this lofty dismissal of philosophy as a useless, irrelevant discipline for finding the truth, because the

---

202 Hawking and Mlodinow, *The Grand Design*, 5.

core discussion of the nature of reality is itself a metaphysical and so a philosophical issue.

The mysterious puzzles behind the metaphysical problem of the origin of the universe and life cannot be solved within a single discipline. Science, philosophy, and theology have to go hand in hand to find the answer to the puzzles behind the mysterious origin of cosmos and life.

PART 4
AN INQUIRY INTO THE MYSTERIES OF SCRIPTURE

CHAPTER 6

WHEN FAITH AND REASON INTERSECT

Mystery creates wonder, and wonder is the basis of man's desire to understand.

Neil Armstrong

WIDENING THE SEARCH FOR THE MISSING PIECE

In my quest for evidence about the origin of the universe and the origin of life, I turned, not only to science but also to the Bible. Could this, different perspective provide missing pieces of the puzzle? Certainly, the Bible does claim to speak with authority about our origins. Christians regard the Bible as "the Word of God"—as the divinely inspired album of the history of the God who created the universe and life. But, because people interpret the Bible in different ways, they come up with different ideas about its meaning and implications. So how should the Bible, especially Genesis chapter 1, be interpreted?

UNLOCKING THE TRUTH OF THE SCRIPTURE

In the first line of the Bible, the foundation of all history is described—"In the beginning, God created the heavens and the earth." Genesis 1:1 makes a bold claim that behind all the mysteries of the universe was the hand of the transcendent, omnipotent God—the creator and sustainer of life. Genesis describes a starting-point for science, for history, and theology.

Genesis 1 elaborates the creation event—when the whole cosmos came into existence at the sound of His voice. Out of nowhere, time, space, and all living things came forth into being as God spoke the universe into existence. With the utterance of His voice, creation, which was formless, takes form; that which was shapeless takes shape; that which was empty fills up with life; chaos yields to order, and light dispels darkness. The word of God has creative power and dynamic effect because it is associated with the voice that speaks. The very first chapter of Genesis shows the power of God's voice. We have this repeated phrase: "And God said . . . " And each time "God said," something happened. That which was formless was shaped into definite form, and that which was empty was filled. The psalmist explains, "at His command they were created." (Psalm 148:5). The great mystery of creation is how something can come from nothing, at this command of God. Paul writes that "God . . . gives life to the dead and calls into being things that were not." (Romans 4:17).

> *Genesis 1:1 makes a bold claim that behind all the mysteries of the universe was the hand of the transcendent, omnipotent God*

In understanding the Bible and nature, the compatibility between them can only be observed if the reader's interpretations are right. Difficulties are bound to arise if there is a wrong scientific interpretation or wrong biblical interpretation. Each book of the Bible was written in its own, particularly historical and cultural context. Unless one puts oneself in the position of the first hearers, one cannot interpret it properly and the real message behind the book, as the author wanted to convey, cannot be grasped. The various authors varied in personality types, cultures, social settings and even differed in their choice of literary genre.

The Bible contains such diverse genres as law, proverbs, poems (such as the Psalms), letters (the epistles) and vision-reports (like Revelation). Some biblical books, like 1 and 2 Kings, are history books—although, of course, none of them can give a complete record of events or they would become unmanageably long (as is mentioned in John 21:25). The first five books of the Bible are, together, regarded by Jews as the Torah, "the law of Moses." But, as well as "laws," they also include a lot of historical narratives, especially about the Exodus from Egypt. Similarly, the book of Genesis is, almost throughout, written in the style of historical narrative, as it gives the records of what happened from the beginning of the world up to the death of Joseph in Egypt. Starting from the story of Creation to the story of Joseph, it is a record of people and events, and so the book belongs within a general framework of biblical history.

Approaches to interpreting the Bible vary widely. On one extreme lies the approach of literalistic interpretation, which involves taking the meaning of the words as much "at face value" as possible. Certainly, it does normally make sense to try to understand the literal meaning of a text, but there are times when this sort of approach breaks down. For example, Jesus' explanation of the meaning of his parable about somebody sowing weeds among the wheat (Matthew 13:24–30; 36–43) shows that his words were not intended to be taken literally—as news or history—but figuratively. The story might not be literally "true," but there is a deeper "truth" contained within it. When somebody reports a dream or vision they have experienced (e.g., Daniel 2, 4, 7–12; Book of Revelation), their report may be accurate, but the meaning is not discovered until the vision is interpreted figuratively rather than literally (e.g., by Joseph, Genesis 40–41). Over-literal interpretation of the idea that "the world cannot be moved" (Ps 93:1; 96:10; 104:5), and of words for the sun "rising" and "going down," led some to interpret the Bible as insisting that the sun goes around the earth, whereas science then showed that the earth goes around the sun. Even scientists today have no difficulty in speaking of "sunrise" and "sunset" because they do not expect people to interpret these ideas so literally.

At the opposite extreme lies the approach of treating the Bible as basically fiction. For example, Richard Dawkins is quoted as saying, "Accounts of Jesus' resurrection and ascension are about as well-documented as Jack and the Beanstalk."[203] Those who take this approach probably accept that the Bible mentions some real places (like Jerusalem), some real people (like the Augustus Caesar) and some real events (like the Persian Empire taking over from the Babylonian Empire), but treat it as an unreliable set of exaggerated or simply invented accounts. Even so, in this approach, biblical texts

---

203 "Richard Dawkins," *Independent*, Dec. 4, 2006.

may still be genuinely ancient; they may reflect oral traditions which go back long before the texts were ever written down. So the biblical texts may still give insight into the cultures and contexts within which they were written. Those who treat the Bible as basically fiction might expect that, over time, other historical and archaeological evidence would gradually build up against the biblical accounts. But, rather, the evidence has tended to show that the Bible is no less and perhaps more reliable than the ancient historians. For example, the Bible mentions Gallio as the Roman Proconsul in Corinth when Paul was there (Acts 18:12). Some doubted the accuracy of this detail since there was no corroborating evidence—until an inscription was unearthed in Delphi which finally confirmed the reference to proconsul Gallio.[204]

When it comes to the account of Creation in Genesis 1, there are some who interpret it in a literalistic way, such as Young Earth Creationists, who treat the account as basically a history. There are others who would describe it as a "myth" and, effectively, treat it as fiction. In between these two extremes lies a further, wide range of approaches to interpretation. In particular, many interpreters would agree that biblical texts of different genres need different approaches to interpretation—an autobiographical, eye-witness account (such as Nehemiah) can be taken more literally than a love poem (Song of Solomon). Each text gives, within itself, some evidence of what kind of writing it is. In Genesis 1 there was no human eye-witness to God's first creative acts so, rather than being history, it seems to be a report of an inspired revelation—perhaps a vision.

UNIFYING SUBJECTIVITY AND OBJECTIVITY

Just as there are problems in interpreting scientific truths, we finite human beings, with all the perils of our ignorance and cultural bias, also face limitations in interpreting Scripture and attempting to discover the truth of God. Interpretations vary, based on people's experience and outlook. The translation and interpretation of Genesis 1 have been controversial because the very first chapter of the very first book of the Bible has a bearing on things people care about—the world and how it came to be. Hence, on the creation account, various models have been

---

204   Dunn, *The Cambridge Companion*, 20.

developed; battle lines have been drawn, and many camps have taken their stance on what happened at the beginning of time.

Comparing Interpretations of Genesis 1

The various interpretations of Genesis 1 that accepts divine authorship can be divided into two major types. The first type accepts that Genesis 1 is a summary of God's creative acts and the second type emphasizes that Genesis 1 is intended to teach us, for the purpose of imparting good theology. Here are the main views of the first type:

Interpretations of Genesis Account as God's Creative Acts

a. *The Young-Earth Creationism*—Young earth creationists believe that God created the universe in six, twenty-four hour days, and conclude that the earth is relatively young. According to this view, the age of the Earth is probably between 6,000 and 10,000 years, if we make calculations based on the genealogies in Genesis. Two of the main proponents of Young Earth Creationism (YEC) were Henry Morris and John C. Whitcomb Jr. They tied their view of the age of the earth to Flood Geology, which is based on the idea that all the sedimentary levels in the earth, including all fossils, were laid down during Noah's flood, and so are quite recent. This view is held by organizations such as the Institute for Creation Research and Answers in Genesis.

b. *The Old-Earth Creationism*— As per this view, the days mentioned in Genesis 1 are not twenty-four hour periods, but unspecified periods of time—ages or epochs. This approach generally accepts the estimated age of the universe as being 13.7 billion years and the estimated age of the earth being 4.56 billion years. Some leading advocates of this position are Christian astrophysicist Hugh Ross and Robert Newman. There are various approaches among the old-earth creationists. The most notable ones are Gap theory and Progressive Creationism. Gap theory states that there is a gap of billions of years between the first verse of Genesis and the second verse and that God recreated the earth after it became formless and void, due to Satan's rebellion. Progressive Creationism (or the Day-age view) states that there were various epochs of time in which God repeatedly and miraculously intervened in nature to create each level of the environment, and its life-forms, before proceeding to the next. The days of Genesis 1 are understood as seven sequential, finite but unspecified, epochs of time.

c. ***Theistic Evolution view***—This view has advocates like C. S. Lewis, Dennis Alexander, and Francis Collins. They suggest that the universe evolved through natural laws pre-programmed by God and that the higher life-forms evolved from lower life-forms through God-guided biological evolution. Most theistic evolutionists believe that Adam is the result of God infusing a pre-existing hominid with a spirit, which implies that the earth is old and that God used evolution in creating the world. They hold the stance that it was through the process of evolution, over a long period of time that God created the diversity of life forms. God created a universe with the potential to develop complex life through natural processes and then did not intervene in that process.

d. ***Intelligent Design Theory***—This theory is an effort to empirically detect whether the "apparent design" in nature is the product of an intelligent cause or is simply the product of an undirected process such as natural selection acting on random variations. The theory holds that natural processes are not sufficient to produce the design in nature and that supernatural intervention is necessary. Among the major proponents of the intelligent design movement are Jonathan Wells, Stephen C. Meyer, John G. West, and Ann Gauger, along with many others who are mostly affiliated with the Discovery Institute, Seatle, USA. They advocate that intelligent design is good science and requires no religious commitment, as there is a scientific basis behind belief in supernatural creation and intervention.

e. ***Revelatory Day view***—Bernard Ramm and P.J. Wiseman, states that the days mentioned in Genesis 1 are not six literal days of creation, but six days of revelation—six days during which God revealed the process by which He had created everything.[205] According to this view, Genesis 1 is a report of this revelation, of a

---

205 Wiseman, *Creation Revealed in Six Days*, 1958.

series of visions of creation given by God, over a period of six days, to Moses (or perhaps to Adam). According to this view, on the first day, God revealed to Moses a vision of the universe with all of the fundamental physical forms of matter that would be transformed into moons, planets, stars, galaxies and everything else, including humankind. The next verse describes the earth after the creation of the universe. So what follows in the seven days is a sequence of visions of creation taking place.

Interpretations of Genesis Account as being a Narrative

The following views approach Genesis as a narrative that instructs the reader, rather than a specific account of creation.

  a. ***The Framework view***—In this view, the creation events are grouped into two triads: the first three creation days consist of forming, where God separates and gathers, and the second three days are days of filling, in which God makes and fills the universe, which was originally formless and empty. According to this view, proposed by Bruce Waltke and Meredith Kline, even if Genesis 1 is a historical account, it is framed around a literary framework to teach theology.

  b. ***The Promised Land view***—This view proposed by John Sailhamer, suggests that Genesis 1:1 describes the creation of the entire cosmos and that Genesis 1:2 to Genesis 2 highlight and detail the way God prepared the Promised Land, which Sailhamer associates with the Garden of Eden.[206] In Genesis 1:2–2:24, the focus is shifted from the whole universe to the Promised Land. So the six days of Genesis 1 are treated as an account of God's process of preparing the Promised Land for humankind.

  c. ***The Cosmic Temple Functionality view***—John Walton suggests that, in Genesis 1, God was not creating things, but forming and giving function to pre-existing things in a process that transformed disorderly creation into an ordered, cosmic temple.[207] The seven days are thus seen as an account of the process of inaugurating the

---

206   Sailhamer, *Genesis Unbound*, 1996.
207   Walton, The Lost World of Genesis, 2009.

cosmic temple. Days 1 to 3 involves three major functions of life—time, weather and food. Days 4 to 6 involve assigning spheres and roles to the cosmic functionaries. God's rest, on Day 7, represents God, from his newly created temple, taking control of the cosmos.

The above approaches may accept that there is an objective truth—that the biblical God created the whole universe and life—but they have used different ways to approach the truth, indicating that there is subjectivity in people's understandings and interpretations. For those interested in the exact age of the universe and of the earth, it is not clear that the Bible is intended to provide answers, just as it is not specific about when the world will end. Jesus says, "No-one knows about that day or hour, not even the angels in heaven, nor the Son, but only the Father."(Mark 13:32). The knowledge of the times and seasons is with the Lord, and yet time and again, mortal men have used their imagination to find out that declared limit on the part of God's revelation. Paul says, "Now we see but a dim reflection as in a mirror; then we shall see face to face. Now I know in part; then I shall know fully, even as I am fully known. Now we see through a glass darkly, but there we shall know to our eternal satisfaction why and how the God has done everything."(1 Cor. 13:12). "No eye has seen, no ear has heard, no heart has imagined, what God has prepared for those who love Him."(1 Cor. 2:9). We can remain thrilled that, if the universe we see around is so breathtakingly beautiful, how much more will the heaven be that the Lord has prepared for us, as Jesus says in John 14:3? Such truths are rational, but may not be amenable to scientific methodology. Science is not the only approach to discovering the truth or to applying rationality; it should be substantiated with the truths from biblical theology.

Genesis 1 is about creation events that happened many years before any written record was made. If we grasp how and why the account was revealed to the author, we can better understand the historical context and the intended meaning of the author while it was written. If we find a contradiction between science and Scripture, we should consider whether we are misinterpreting one or the other, because as humans, we are not infallible interpreters, even if we consider God's word to be infallible. In my view, Scripture does describe the special creation of mankind, but it should not be seen as specifying the age of the earth. The latest scientific evidence suggests "old" earth, but current theories, based on evolution, about the origin of life and of species are not coherent and have some very significant limitations.

CONCLUSION

Some dismiss Genesis as fiction which should not be taken seriously, while others treat it as a rigorously accurate history, every detail of which should be taken literally. But there are many who both take Genesis seriously and treat it as needing careful interpretation. If you want to understand Genesis, it makes sense to read it in the light of the original language and culture of this ancient text. At the same time, it is important to appreciate that the events and processes described are actually beyond the understanding and imagination, not only of the author of Genesis, but even of us, still today. The aim of the next chapter is to propose a particular approach to that kind of careful interpretation of Genesis chapter 1.

CHAPTER 7

THE REVELATION MODEL

IT'S FOR A REASON

During a 16-hour flight, I found myself working on this book, while the young man sitting next to me took out a few books to read. We eventually struck up a conversation. Benje was from San Francisco, and he had been working in Pakistan for a few years. We spoke about food and other general topics until I told him about the book I was working on. I explained that I was on a journey to find the truth behind the origin of the universe and life and that my findings led me to believe that the first chapter of Genesis closely agrees with scientific truth. By this time, he had put all his books aside and was curious to know more about my findings and how this was all possible.

"Do you think that the whole universe was created in seven days? Is that consistent with science?"

"Where did you learn that the whole universe was created in seven days? Is that from the Bible?" I asked in return.

"That must be from the Bible. I went to a Christian school where they taught us that the whole universe and all the life we see came into existence in just seven days. You know, they never even allowed us to learn about evolution. I am no longer a Christian, by the way."

"Oh... If you don't mind, can you tell me what made you stop being a Christian?"

"Well, there were many reasons, but this one that we are discussing, that the Bible is anti-science, is one of the main reasons. When a holy book that is said to be from God cannot be trusted, how can that God be trusted?"

> *Benje was just one of the many people out there who had no idea that the Bible might be scientifically accurate*

"That is a good point, Benje. But I'm not sure that the Bible is anti-science. It is undoubtedly advocating the idea that there was nothing before and everything came into existence because of God, the supernatural being. But is the Bible insisting that a 24-hour day, as we experience now, is exactly the kind of day that God was experiencing before there were any people—before there was even a sun to rise and set? Though the Bible is not a science book, the scientific information given in the Bible does seem to be right, for example, the sequence of Creation . . . "

So, the conversation went on from there to many other topics on science and faith. This young man and his questions had an impact on me. Although I had begun work on the book because of an idea that interested me, I now had a greater drive and reason to delve deeper into the biblical account of creation and how it coincides with scientific evidence, because Benje was just one of the many people out there who had no idea that the Bible might be scientifically accurate.

INTERPRETING GENESIS CHAPTER 1

Genesis 1 has been the source of controversy over the issue of the origin of the universe and life, along with various problems with the interpretations of scientific evidence, complicated by limited information, cultural bias and so on. The events that happened before humanity even existed, and the events that will happen after humanity ceases to exist, had to be communicated supernaturally through a divine agent, or visions or dreams. The Revelation, the final book of the Bible, is about the events that will happen at the end of the world, revealed through visions, and I concluded that the first chapter of the Bible must also have come through a process of divine revelation.

According to the Revelatory model proposed in this book, creation didn't happen in six 24-hour periods, but it was communicated, through visions and dreams, over six consecutive days. While Genesis 1 should not be dismissed as mere fiction, it does not fit within the genre of "history" either, since there was no possibility of a human eyewitness for most of the process of creation. It is a description of the creation from God's point of view. The reader or hearer of Genesis is led to assume that this account was communicated by God to its human author (or authors). Throughout history, Jews have tended to assume that Moses was the author of Genesis, and indeed of all five books of the *Torah*, because he met God "face to face" (Exodus 33:11) and was told to write down things that God told him (Exodus 17:14; 24:4; 34:27).

The "Days" of Creation

The account of creation can be viewed as a revelation by God to the human author. So the "days" of Genesis 1 are understood, in these chapters, as the structure used to describe creation—perhaps the process of revelation was spread out over seven sequential, finite days. But the creation itself is understood as happening in epochs of time, and the exact length of these epochs is undetermined. There are several reasons for proposing this interpretation of "day" as "epoch," rather than as a 24-hour-period.

First, it would be an anachronism to assume that the first "day" consisted of 24 hours—at a time when there was not yet any earth to spin or any sun to rise or set. It is these cycles which set the length of a literal, 24-hour day (although, technically, our measurements of time have now been redefined in terms of the radiation from Caesium atoms).

Secondly, the Bible makes it clear that God's perspective on time is not necessarily the same as ours (Psalm 90:4; 2 Peter 3:8), allowing 1,000 years to seem like a day; and the early verses of Genesis must be described from God's point of view, because there were not yet any humans to watch creation happening.

Thirdly, the Hebrew Scriptures in general, including the book of Genesis, allow the word "day" a rather wider range of meanings than simply a 24-hour-period. It is used about time more generally (e.g., Genesis 4:3). In the same way, we might say, "That'll be the day"— meaning something more like "moment" or "time."

Finally, the use of "day" here allows a description of the original events (of creation) to be connected to a routine of remembering them—with the Jews resting every Sabbath, or every seventh day and one reason for this is to echo, celebrate and remember what God did (Genesis 2:2–3 echoed in Exodus 20:11). The Sabbath rest certainly does involve a literal, 24-hour day. But does that require those who believe the Bible to interpret it as saying that the creation must have happened in six 24-hour periods? The Bible includes another similar, way of celebrating and remembering which suggests not. The Exodus of the Israelites from Egypt was celebrated every year with Passover, on one specific day, the 14th of Abib (Exodus 12:6; Deuteronomy 16:10). The purpose of this celebration is "that you may remember the day of your departure from Egypt" (Deut. 16:3). But the departure being celebrated and remembered, although described as a "day," lasted much more than 24-hours. The Passover meal was eaten on the evening of the 14th of Abib, but the departure began only the next day (Numbers 33:3) and involved a slow and lengthy journey for more than 600,000 men, armed for battle (Exodus 13:18), with their families, flocks, and herds (Exodus 12:32), which cannot travel very fast. They must have started near the Nile, and they only reached "the edge of the desert" at their second encampment (Numbers 33:6). Then it was only after they crossed through the Sea of Reeds (Red Sea) that they were finally out of the "land" (Exodus 13:3) of Egypt.

The Jewish celebration of the Passover was to be on a specific day, but it does not require the belief that the Exodus from Egypt must be taken as literally happening within 24 hours. In the same way, Jewish observance of rest every Sabbath, strictly keeping every seventh day "holy," involves remembering God's process of creation (Genesis 2:2–3), but it is not clear that it requires the belief that creation happened in literally and exactly six 24-hour days.

The creation started with—an earth without form and void and ended with an earth that is full of all kinds of life, including the highest form of life—the human being, made in God's very own image. God showed these creation events to Moses in the correct order, and Moses recorded it as he saw or heard it. There is an in-built chronology of physical creation events in Genesis 1. Some might object that the

Genesis account is not complete—for example, it mentions plants on land (Genesis 1:11–12) but omits to mention plants in the water, and it is not quite clear where amphibians or dinosaurs would fit in. Indeed, the account is only a summary—further details are mentioned elsewhere in the Bible, e.g., God forming mountains (Amos 4:13); the presumably figurative ideas of God creating with his hands (Isaiah 45:12) or fingers (Psalm 8:3) or with a womb (Job 38:8,29); the angels watching creation happen, and shouting with joy (Job 38:7); and monstrous beasts as among the creatures God made (Great whales in Genesis 1:21; land monster "behemoth" in Job 40:15–24; sea-monster in Psalm 74:13; "leviathan" in Isaiah 51:9, and "great monster" in Ezekiel 29:3). So the Bible itself does not treat Genesis as a complete account. The approach used here treats the first book of the Bible (Genesis) as similar to the last book (Revelation)—they are both revelations of God to His prophets. While Genesis 1 reveals the history of the creation of the universe, Revelation portrays human history's culmination.

The phrase "and there was evening, and there was morning" is used for six days but is not present on the seventh day, to show that the seventh day is different from the other days. It reveals that, though the works of God in creation are completed and will not be repeated, He is still working on redemption and the formation of the new heaven and new earth. The act of God resting does not mean that God was tired, but arises because the seventh day of the creation week was a prototype for (or a "type" of) the rest associated with the Jewish Sabbath (Exodus 20:11). This completion of creation can be understood with the majestic beginning of Genesis 1:1 with the word "beginning" and the account ending with the word "finished" in Genesis 2:1–2.

With Genesis 1:1, we have a beautiful introduction to the creation account, and with Genesis 2:2, it culminates with a wonderful ending. Creation starts with a world that is without form and void and ends with a world that is full of all kinds of life, including the highest form of life—the human being. There is an inbuilt chronology of natural history—a chronology of physical creation events in Genesis 1 and so, anyone who reads it for the first time may naturally interpret it as a chronology of physical creation events. God created all of the creation sequentially with a goal, the goal of preparing the earth for His special creation, human beings, in His very own image, who were created to enjoy fellowship with God.

> *I find the Biblical details of Creation to be in harmony with the latest cosmological evidence and observations*

I find the biblical details of Creation to be in harmony with the latest cosmological evidence and observations, starting from the initial cosmic beginning, to its expansion and the formation of stars and galaxies and finally the earth, which God specially made for us. As astronomical research advanced, it proved some aspects of the biblical account. The atheistic astrophysicist Fred Hoyle concluded, "There is a good deal of cosmology in the Bible. . . . It is a remarkable conception."

In the following chapters, I would like to take you on a journey to show you how, whatever the lengths of the various epochs involved, the sequence of creation beautifully matches the latest scientific, cosmological findings.

CONCLUSION

The events described at the beginning of Genesis 1 happened before there was any human to witness them, so they are clearly described from God's point of view. The rest of the Bible treats this account, not as fiction, but as a reliable account. That presupposes that these details must have been communicated somehow by God to the human author—perhaps Moses. This book concludes that, just as visions are the way the future is communicated in the book of Revelation, similarly the account of creation was revealed through a series of visions—perhaps on six consecutive days—which tell of the stages, or epochs, of creation in a fascinating sequence. The next chapter shows how this sequence, presented in Genesis, can be related to the scientific evidence.

CHAPTER 8

THE FASCINATING SEQUENCE

IN THE BEGINNING?

The story of the cosmos is exciting and thought provoking. This chapter will navigate you through the story in a chronological sequence, from the beginning of the universe (space, time, and energy) to the appearance of humans. To present the complete story, the two disparate descriptions of reality—science and the Bible—are both necessary and, in the next few pages, you will see those two realities beautifully interwoven while the story unfolds. I urge you to give careful thought to this exciting story as some parts may be puzzling, but I can assure you that, as you journey until the end, you will explore the very history of time and space, the dawn of light, the origin of earth and the origin of our forefathers. Be ready as you are about to embark on a wonderful journey to traverse through the various epochs of the universe.

The idea that the universe started in a 'Big Bang' is accepted by most of the cosmologists, and hundreds of books and journals are available on this idea, some of them providing strong evidence to substantiate this model. But there is no consensus on how the Big Bang originated, for which the Bible gives a beautiful answer—*In the beginning, God created the heavens and the earth (Gen 1:1)*—the Big Bang originated from God, who is the uncaused cause and the timeless Being. Many do not want to fit anything supernatural into the story of the origin of the universe, but existing scientific theories do not provide a complete or adequate answer, and there is no inclusive theory to account for the origin of the universe if the supernatural is excluded.

Without a supernatural Being, the questions about the origin remain: How did the Big Bang come into existence—from nothing? Or some pre-existing matter? Was this event a miracle, falling beyond the boundaries of science? The great scientist Stephen Hawking had similar questions, as he did not at any cost want to acknowledge the involvement of the supernatural. He wrote, "We find ourselves in a bewildering world. We want to make sense of what we see around us and to ask: What is the nature of the universe? What is our place in it and where did it and we come from? Why is it the way it is?"[208]

> Without a supernatural Being, the questions about the origin remain: How did the Big Bang come into existence—from nothing?

The second part of Genesis 1:1 says that *"Now the earth was formless and empty, darkness was over the surface of the deep, and the Spirit of God was hovering over the waters."* The Bible does not explain exactly how long the universe was plunged in this darkness or how long it remained empty. But modern cosmology gives us some indications about our origins.

The Early Universe: The First Three Minutes

The universe was unimaginably hot and dense when it was born. Whatever one might imagine from the term Big Bang, space did not simply expand from one spot, but space appeared simultaneously everywhere in the universe with the Big Bang; space stretched and carried matter with it. According to the inflation theory, when the universe was just 10^{-34} of a second or so, it went through inflation; an incredible spurt of expansion, in which the universe expanded from a tiny point to a ball sized universe much faster than the speed of light instantaneously. Weinberg gives a brief account of this period—At about one-hundredth of a second, the temperature of the universe was about $10^{11}\,°C$, much hotter than the hottest star and, the matter that was rushing apart consisted only of elementary particles.[209] Calculation

---

208   Hawking, *A Brief History of Time*, 171.
209   Weinberg, *The First Three Minutes, 1993*.

shows that the density of this cosmic soup at a temperature of 10^{11} °C was about 4×10^9 times the density of water! It was so hot that none of the components of ordinary matter—molecules, or atoms, or even the nuclei of atoms could have held together.

> Cosmologists suspect that the four forces that rule the universe—gravity, electromagnetism and the weak and strong nuclear forces—were unified into a single force at the universe's birth, squashed together because of the extreme temperatures and densities involved. But things changed as the universe expanded and cooled. Around the time of inflation, the strong force likely separated out. And by about 10 trillionths of a second after the Big Bang, the electromagnetic and weak forces became distinct, too. Just after inflation, the universe was likely filled with hot, dense plasma. But by around 1 microsecond (10 to the minus 6 seconds) or so, it had cooled enough to allow the first protons and neutrons to form—researchers think. In the first three minutes after the Big Bang, these protons and neutrons began fusing together, forming deuterium (also known as heavy hydrogen). Deuterium atoms then joined up with each other, forming helium-4.[210]

The resulting gas, with Helium, Hydrogen, and Deuterium, condensed finally to form the galaxies and stars. The first three minutes were responsible for the ingredients to form the first stars.

The Dark Universe: From Three Minutes Onwards

"Because the universe has the conditions of the core of a star, it had the temperature and pressure to fuse hydrogen into helium and other heavier elements. Based on the ratio of those elements we see in the universe today: 74 % hydrogen, 25 % helium and 1% miscellaneous, we know how long the universe was in this "whole universe is a star" condition. It lasted about 17 minutes from 3 minutes after the Big Bang until about 20 minutes after the Big Bang."[211] Since everything was hot and dense, filled with electrically charged ions with the matter distributed as highly ionized plasma, the photons (light particles) could not travel, and there was no light. Everything was "formless," and there was "darkness" (Genesis 1:1), that lasted for several hundred

---

210 Wall, "The Big Bang" *Space*, Oct. 2011.
211 Cain, "When was the first light," *Physics*, Nov. 2016.

million years. This era that existed before the formation of the first stars and galaxies is a time that is devoid of data, mostly remaining a mystery, stopping us from observing the state of the primordial universe.

DAY 1: LET THERE BE LIGHT

Genesis 1:3–5 And God said, "Let there be light," and there was light. God saw that the light was good, and he separated the light from the darkness. God called the light "day," and the darkness he called "night." And there was evening, and there was morning—the first day.

The first step that the infant universe took after its birth was to come out of the darkness that it was immersed in, with no detectable light even though the photons were present. The supernatural power of the Lord divided the light from the darkness, as the Spirit of God brooded over whatever was in existence before light emerged.

The Era of Recombination

As the universe expanded, it started to cool down, and about 380,000 years after the Big Bang, the ions and electrons could "recombine" to form neutral atoms of hydrogen and helium. As the ionised plasma gave way to neutral atoms, the photons were not scattered but could travel freely and, the universe went from being opaque to transparent. This was the moment of the first light in the universe known as the "Era of Recombination." All the photons emitted since this time is the reason for the cosmic microwave background radiation that allows astronomers to study objects all the way back to recombination. "Because the universe has been expanding over the 13.8 billion years from then until now, those earliest photons were stretched out, or red-shifted, from ultraviolet and visible light into the microwave end of the spectrum. If you could see the universe with microwave eyes, you'd see that first blast of radiation in all directions."[212]

If we would like to see our internal organs, we have to undergo scanning, and similarly, the picture of the baby universe can be seen through the remnant of the Big Bang—the cosmic microwave

---

212   Cain, "When was the first light," *Physics*, Nov. 2016

background radiation that can be seen as a glowing imprint on the entire sky. The measurement of this radiation through increasingly powerful telescopes has enabled astronomers to peer into the early universe with unprecedented accuracy. Allowing for the time light takes to travel to the telescopes, astronomers can capture snapshots of the infant universe and attempt to calculate its age.

The Dark Age

After that first blast of light, during the period of recombination, everything was dark, filled with enormous amounts of primordial elements, with no stars or galaxies. "As a result, the universe entered a period known as the dark ages that span the time between recombination and the formation of the first stars. Just how long that Dark Age lasted has been difficult to pin down. Based on observations such as the Hubble Deep Field, it's been estimated that the dark ages ended about 400 million years after the big bang."[213] As per cosmologists, at the beginning of these dark ages, the temperature of the entire universe was about 3726 °C (4000 Kelvin), and the temperature went down to 100 °C (373 K), the boiling point of water 10 million years after the big bang. And then 7 million years later, it was down to 0°C (273 K), the freezing point of water and finally dropped to 60 Kelvin by the time Dark Age ended.

The Cosmic Dawn: Epoch of Reionization

At the end of the Dark Age, cold primordial elements like hydrogen and helium came together, and the first stars were formed. They are the Population III stars, with hundreds of times the mass of our sun, and these first stars had no heavy elements. Some cosmological models show that they "had surface temperatures of about 100,000 Kelvins—about 17 times higher than the sun's surface temperature. Therefore, the first starlight in the universe would have been mainly ultraviolet radiation from very hot stars, and it would have begun to heat and ionize the neutral hydrogen and helium gas around these stars soon after they formed. We call this event the cosmic renaissance."[214] Though

---

213   Koberlein, "Second Light," *Cosmology*, Feb. 2015.
214   Larson, "The First Stars," *Scientific American*, Jan. 2009.

there is no direct evidence for the very first hot stars, scientists believe that they turned on around 180 million years after the Big Bang, lighting the universe with the very first light.

However, these first stars had short lives and, at the end of their brief lives, while some exploded as supernovae, others collapsed into massive black holes and became the power sources for quasars. As the stars exploded, they seeded the Universe with all the chemical elements that made the next generation (Population II) stars, which include heavier elements, to form. They too exploded as supernovae, and the whole universe lit up, with stars going off violently like fireworks. Thus the neutral universe became reionized ushering in the Epoch of Reionization. The heavier elements, dispersed into intergalactic space by the exploding stars, contaminated other collapsing proto-galactic clumps (the first star-forming systems), which coalesced over time into the larger galaxies of today. The youngest stars, called Population I stars, are the metal-rich stars and our sun is one of them. The metal-rich stars and galaxies which we can see today are the result of supernovae explosions of those very first stars that lit the whole universe many billion years ago.

Genesis says "and there was light." Our scientific understanding agrees that the universe seems to have emerged suddenly out of nothing. Light (photons) may have been a part of the Big Bang, but light we could observe only came once stars began to light up the universe. Science agrees that light was a key part of the creation of the universe and it was the very first thing that emerged.

DAY 2: LET THERE BE A HABITABLE PLANET

Genesis 1:6–8 And God said, "Let there be a vault between the waters to separate water from water." So God made the vault and separated the water under the vault from the water above it. And it was so. God called the vault "sky." And there was evening, and there was morning—the second day.

After creating light, and separating it from darkness, God started preparing a habitable planet where life can thrive. God reveals, on "day 2", how He made the sky, which is described as a "vault" ("firmament" in some old translations), and how He provided separate layers for protection. This means that God was preparing the space around the earth so that our planet can have a habitable environment. This involved two major processes—positioning the planet in the right place and, creating protective (defense) layers. Scientists have identified over 150 parameters for life to exist within our solar system, and each parameter is so exacting that they could not have happened by chance. So Genesis provides a step-by-step summary of how the universe came to be fine-tuned for life on earth.

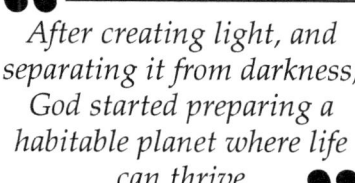

After creating light, and separating it from darkness, God started preparing a habitable planet where life can thrive

Positioning in the Goldilocks Zone

The "Goldilocks Zone" is the habitable zone with just the right temperature, where life can exist on a planet. This involves positioning our sun in the right place within the Milky Way galaxy and then positioning our earth in the right place within our solar system.

Positioning our Sun inside the Milky Way Galaxy

As many of the stars came to the end of their lives, the heavy elements in the dust and gas they spewed around them drew together and took the shape of the billions of galaxies we see around us, including our home. As we observe the whole universe from our vantage point on earth or a space telescope, one can understand how difficult it is to count the number of galaxies or to find the limit of this huge universe. Galaxies exist in clusters, which are grouped into superclusters measuring hundreds of millions of light-years across. Galaxies can be classified into three major types, according to Hubble's classification scheme: elliptical, spiral and irregular, based on their shapes. Elliptical galaxies are some of the oldest and most massive galaxies in the cosmos and usually have super-massive black holes. Spiral galaxies, such as our own Milky Way, also known as disk galaxies, are highly flattened and are less bright than elliptical. Galaxies that do not fit into the spiral or elliptical classifications are called Irregular galaxies.

Our home galaxy—the Milky Way was formed about 10 billion years ago. It is a spiral galaxy, having around 100 to 400 billion stars in the form of a thin disk around 100,000 light-years in diameter and just a few thousand light-years in thickness. The Milky Way galaxy has four major spiral arms, and our sun is located in the inner rim of a minor arm, or spur called the Orion arm, between two of the major arms. The positioning of our solar system close to the inner rim of the minor arm of the Milky Way galaxy is highly strategic because the major arms consists of supergiant stars, surrounded by intensely hot plasma where the star densities are high and, less suitable for life. Because of our positioning between the spiral arms, we can get a clear-viewed cosmic window that allows us to see the other parts of our galaxy and also beyond, to the hundreds of billions of other galaxies across the universe.

> *The positioning of our solar system close to the inner rim of the minor arm of the Milky Way galaxy is highly strategic*

Positioning our Planet inside our Solar System

Our sun was formed, like all stars, by the gravitational collapse of a giant molecular cloud of gas and dust grains, called the solar Nebula, about 4.6 billion years ago.[215] "As the Sun formed in the center of the disk-shaped Nebula using around 99 percent of the material, the remaining material began to clump together, forming planets and dwarf planets. The solar wind swept away lighter elements, such as hydrogen and helium, from the closer regions, leaving only heavy, rocky materials and thus terrestrial planets, like Mercury, Venus, Earth, and Mars formed near the Sun."[216] The four planets that formed farthest from the Sun are the Jovian planets: Jupiter, Saturn, Uranus, and Neptune, which are essentially big balls of gas, surrounded by many moons and rings. The ice-infused bits and pieces of the early solar system that could never form a planet became the asteroids in the Asteroid belt, while some icy and gaseous material coalesced into comets in two particular regions of the solar system: the Kuiper belt and the Oort cloud. Pluto is a rocky ice planet (dwarf), and the largest of the Kuiper belt objects.

---

215 Redd, "How Was Earth," *Space*, Oct. 2016.
216 Redd, "How Did Solar System," *Space*, Jan. 2017.

The dividing line between Jovian planets and the terrestrial planets is called the frost line that defines the boundary beyond which hydrogen compounds like methane (CH_4), and ammonia (NH_3) are simply frozen like ice. Our home planet is situated within that boundary, in the habitable zone, where the temperature is warm enough to allow for abundant liquid water. "If our planet were just a little closer to the Sun, a runaway greenhouse effect would render it unliveable, a climate model suggests . . . if the planet were to receive about one-tenth more solar radiation than it does now, that blessing would become a curse: water vapor traps heat just like atmospheric carbon dioxide, and on a warmer world more water would evaporate, boosting the greenhouse effect and thus trapping even more heat. That destabilizing feedback, or runaway greenhouse effect, would eventually rob the planet of its oceans."[217]

Creating Protective Layers

The Heliosphere

The great "vault" refers to the expanse of the sky that is made up of separate layers. The first layer of defense of our home planet from galactic cosmic rays coming from outside the solar system is provided by a bubble of solar magnetism called "the heliosphere," which encircles all the planets up to Pluto and beyond. "This springs from the sun's inner magnetic dynamo and is inflated to enormous proportions by the solar wind. When a cosmic ray tries to enter the solar system, it must fight through the heliosphere's outer layers; and if it makes it inside, there is a thicket of magnetic fields waiting to scatter and deflect the intruder. Once a cosmic ray enters the solar system, it must "swim upstream" against the solar wind."[218]

The Earth's Atmosphere

The earth's atmosphere is a 500 km thick layer composed of a mixture of gases; it is a mix of about 78 percent nitrogen, 21 percent oxygen, with small amounts of other gases, especially carbon dioxide. Carbon dioxide, water vapor, methane, ozone, and nitrous oxide constitute

---

217 Perkins, "Earth is just within," *Nature*, Dec. 2013.
218 Phillips, "Cosmic Rays Hit." *NASA*.

greenhouse gases as "they act as a partial blanket for the thermal radiation from the surface and enable it to be substantially warmer than it would otherwise be, analogous to the effect of a greenhouse. This blanketing is known as the natural greenhouse effect. Without the greenhouse gases, Earth's average temperature would be around -20 °C."[219] The greenhouse gases have a huge effect on climate, even though our Earth's atmosphere has a tiny amount of them, as they allow the earth's surface to be warm by trapping heat in the atmosphere, preventing the temperature not to reduce below freezing. The more greenhouse gases, the more heat will be retained, warming our planet.

The atmospheres of our two neighboring planets, Venus and Mars, are very different—mainly carbon dioxide. The atmosphere on Mars is extremely thin, with high winds. Jupiter, Saturn, Uranus, and Neptune have very dense atmospheres, with hydrogen, helium, methane, and ammonia. Mercury has no atmosphere at all. Our atmosphere on earth changes with height in its density, temperature, and composition. It can be divided into various layers:

a. *The Troposphere* is the lowest layer, containing half of the atmosphere.

b. *The Stratosphere* contains about 90 percent of the ozone in the atmosphere, and this ozone layer absorbs most harmful ultraviolet rays, known as UV-C and UV-B, from the Sun, allowing life to exist. Ultraviolet rays can damage plant tissue. In humans, this radiation causes such problems as skin cancer and tissue damage to the eyes.

c. *The Mesosphere* protects us from meteors, as meteors burn up due to friction with the gas molecules and the temperature of this layer reaches its coldest (~ -90 °C).

d. *The Thermosphere* can get heated up to 1,500 °C or higher and can cause molecules to ionize.

e. *The Ionosphere*, filled with charged particles, making long-distance radio communication possible, is a part of Thermosphere. Above the ionosphere lies the magnetosphere, the giant magnetic bubble

---

219   http://www.theozonehole.com/atmosphere.htm.

surrounding the earth. Particles coming from the sun, carried in the "solar wind," are prevented by this layer from actually hitting the earth.

f. In *the Exosphere,* the atmosphere merges into space.

Also, the positioning of Jupiter is such that comets are unlikely to hit our planet because a comet has to get past Jupiter before it can crash into Earth. Comets that come too close to the great mass of Jupiter tend to be flung out of the solar system because of its gravity.

After creating the earth, before going on to populate it with life, God positioned it in the right place and made various layers of the sky ("vault") to make our earth habitable. Science fills in more details about how planet earth formed and shows how many details had to be exactly right for this planet to be habitable. The positioning and protection of our home planet cause even skeptics to pause and wonder whether there is a super intelligence behind all this.

> *After creating the earth, before going on to populate it with life, God positioned it in the right place*

DAY 3: LET THERE BE LIFE

Genesis 1:9–13 And God said, "Let the water under the sky be gathered to one place, and let dry ground appear." And it was so. God called the dry ground "land," and the gathered waters he called "seas." And God saw that it was good. Then God said, "Let the land produce vegetation: seed-bearing plants and trees on the land that bear fruit with seed in it, according to their various kinds." And it was so. The land produced vegetation: plants bearing seed according to their kinds and trees bearing fruit with seed in it according to their kinds. And God saw that it was good. And there was evening, and there was morning—the third day.

After positioning our home and creating proper defenses for it, the earth was ready for the amazing step of the emergence of first life. Dr. John Lennox says that "the yawning gulf between inorganic and organic matter is underlined in Genesis by the fact that on day 3 God spoke twice."[220] However, the basic ingredient needed for life, water was not available on planet earth when it was formed. On the third day, God revealed how he first brought water, made the mighty ocean, separated it from the land, and then went on to make the first organic molecules—life in the form of microscopic organisms and plants. That brings us to the story of the origin of the mighty ocean, where the first life appeared.

The Origin of Oceans

I have always been fascinated by the ocean, and I believe that there is something mysterious about it. Every time I walk along a beach, my emotion and mood are connected to the waves; they stir my heart and inspire my imagination. I feel as though the voice of the sea is continuously trying to talk to my soul and I try to listen by moving closer to its waves. That is why William Wordsworth said that the ocean is a mighty harmonist. However, the origin of oceans itself is a mystery, as still now scientists are not sure how water came to these mighty oceans. First, water was somehow brought to our planet and, second, a boundary was somehow established for that water, so that the land was separated from the ocean, enabling life to exist both on the land and in the water.

The Mystery of the Origin of Earth's Water

When our solar system formed, the oceans' signature ingredient, water (97 percent of earth's water is found in oceans) was not found due to the blazing heat of the sun and the earth's relatively feeble gravity couldn't hold the water vapor. The latest research points to an out-of-this-world arrival of water to our home planet. The origin of our planet's water is an intricate story and research published in an article points to two possible sources[221]—

    a. Comets—Some scientists suspect that water was brought by comets that pummeled our earth during its early stages. In recent years, this hypothesis has lost favor because ESA's Rosetta probe, along with some earlier missions to Halley's Comet, showed that the chemical composition of the water on comets does not match

---

220   Lennox, *Seven Days that divide*, 70.
221   Crockett, "How did Earth get" *Science News*, May 2015.

the water on earth. Water consists of two hydrogen atoms and a single oxygen atom. Sometimes deuterium sneaks its way into the mix. When one or both hydrogen atoms in water is replaced by an extra neutron (2H, rather than 1H), we get the heavier version of water, deuterium. Water on earth only has about 156 deuterium molecules in a million molecules of water. Research has shown that typically Comets from the Oort cloud and the Kuiper belt have double the deuterium to hydrogen or D/H ratio, compared with water on Earth. In 2014, Rosetta sampled the water from Comet 67P and found surprisingly high levels of deuterium—the D/H ratio was more than three times that of water on earth.

b. *Asteroids*—The asteroid belt makes another potential source of water. During the initial formation period of our solar system, some of the icy asteroids formed beyond the Frost Line would have been flung inwards towards the inner rocky planets, including Earth. Some of these asteroids contain as much as 20 percent water. Computer simulations have suggested that, from 8 million years after the solar system formed, a hail of asteroids might have delivered ice in large quantities to account for our oceans. Researchers found that the D/H ratio of the water trapped in meteorites from the asteroid Vesta, one of the largest asteroids between Mars and Jupiter, closely matches the composition of Earth's oceans. However, the exact source that delivered water to our planet is still a mystery.

Another mysterious fact is that though the water delivered by asteroids or meteorites would have been delivered to all the inner planets, only the planet earth was able to sustain that water. Photographs of the surface of both Mars and Moon show dried river beds. On Mars, NASA's Curiosity Mars Rover discovered an ancient streambed, suggesting that there was once water there, a long time ago. But the charged particles of the solar wind removed what atmosphere Mars once had, leaving the planet dry.

Drawing the Boundary Line—Origins of Continents

Our planet consists of five major oceans—the Pacific, Atlantic, Indian, Arctic, and Southern oceans, and the continents that frame them define the boundaries of each ocean. However, if we journey back and visit the earth many millions of years ago, we would have been shocked to see a different planet. Nearly four billion years ago, our planet was mostly covered by oceans and almost all the land that was there was lying along the Equator constituting a single supercontinent called Pangaea. Plate Tectonic Theory, which is accepted by most geologists, says that the pieces of Pangaea that began to move apart became the modern continents. According to this theory, the top two layers of earth (the crust and the top portion of mantle) is broken up into sections called tectonic plates and slides on the molten mantle due to the heat from the core of the earth, and this interaction called plate tectonics contributed to the creation of continents. First two billion years of Earth's 4.6-billion-year history were dominated by volcanic activity that generated the crust which is thick and rich in silica; then liquid water collected on its surface, forming oceans which later got divided into continents around 3 billion years ago. The positions of the continents are still changing and who knows, if we probably visit our home after millions of years later, the US state of California would have become an island, and Africa might have been split in two.

Earth is the only known planet whose surface is divided into continents and oceans. The Bible suggests that it was God who determined the boundaries of the oceans, as suggested in Job 38:8–11. "Who enclosed the sea with doors [shorelines] when it burst forth and went out of the womb; when I made the clouds its garment and thick darkness its swaddling band, and marked for it My boundary and set bars and doors and said, 'This far you shall come, but no farther; and here your proud waves shall stop'?"

In the Bible's sequence, after making the universe, God makes a habitable planet, with both land and sea, before placing life—in the form of plants—on this habitable planet. This agrees with the sequence suggested by science.

The Origin of Plants

After forming the oceans, and separating the land from the sea, God went on to create plants. "The Earth's history spans 4.6 billion years! Geologists have created a time scale to organize Earth's history into eons, eras, periods, and epochs."[222] The oldest or first era, between 4.6 billion years to 542 million years ago, where very few fossils are found is called the Precambrian era or "obscure life" era. This era refers to earth's entire history up to the point when the first rocks containing recognizable fossils were formed. The succeeding eras are—the Paleozoic Era (Permian, Pennsylvanian, Mississippian, Denovian, Silurian, Ordovician and Cambrian period), the Mesozoic Era (Cretaceous, Jurassic, and Triassic period), and the Cenozoic Era (Quaternary and Tertiary period).

There are huge varieties of life on earth and all "organisms are classified into three Domains and into one of six Kingdoms of life. These Kingdoms are Archaebacteria, Eubacteria, Protista, Fungi, Plantae, and Animalia . . . Archaebacteria are single-celled prokaryotes originally thought to be bacteria...live in some very inhospitable places, such as hot springs and hydrothermal vents . . . Eubacteria are considered to be true bacteria . . . Fungi include both unicellular (yeast and molds) and multicellular (mushrooms) organisms . . . They decompose organic matter and acquire nutrients through absorption."[223] All microscopic organisms that do not belong to bacteria, plants or fungi are Protists such as the slime molds and algae. Animals are divided into vertebrates (fishes, amphibians, reptiles, birds, and mammals that have a vertebral column) and invertebrates. The invertebrates that are characterized by their lack of backbones account for 97 percent of all animal species appeared first and include insects, worms, arthropods, sponges, mollusks, octopuses, and countless other families.

The Bible does not give an account of all the Kingdoms. The Protists, Fungi, Prokaryotes and invertebrates, like insects and worms, that aid in pollination could have been created along with the plants or immediately after the plants. God orchestrated the Creation sequence so beautifully and created plants of various varieties at different instants of time, after making the environment conducive and right for their existence.

> *God orchestrated the Creation sequence so beautifully*

---

222 Geology, "Geologic Time Scale."
223 Bailey, "The 6 Kingdoms of Life." 2018.

Plants couldn't have come into existence until there was a sufficient ozone layer to block radiation. The Paleozoic era from 542 million years ago to 250 million years, began with the Cambrian Explosion where suddenly life started to flourish on Earth. Scientists say that plants and invertebrates were seen first. During the Paleozoic Era, Protists, Fungi, Archaebacteria, Eubacteria, invertebrates, and plants appeared. Scientists at the Swedish Museum of Natural History have reported that they have found fossils of 1.6 billion-year-old probable red algae.[224]

The fossil record shows that the major groups of plants appear suddenly and fully formed around 400 million years ago. The seed plants include the gymnosperms (whose seeds are not enclosed in an ovule), and plants like conifers, ginkgo, and cycads that had an adaptive advantage and dominated the forests of the late Paleozoic era. The first fossils of vascular plants have been seen at 375 million years ago. The flowering plants appeared in the fossil record fully developed and the fossils show no evidence of changes that might have occurred in developing from the lower forms to the flowering plants. There are no known ancestors for a majority of the major phyla of plants.

God created the herbs and plants first before creating the other living forms, because these green plants helped in the removal of carbon dioxide from the air, and replace it with oxygen. The early days of Earth contained only 5 to 10 percent oxygen. Over the years, while carbon-dioxide decreased, the percentage of oxygen in the air gradually increased because of plants. The increase of oxygen is also due to algae, as oxygen is released as they die, sink to the bottom of the ocean and are caused to decompose by bacteria. "Increasing amounts of oxygen in the atmosphere could have triggered the first of three past episodes when the Earth became a giant snowball, covered from pole to pole by ice and frozen oceans."[225] Thus, "by simultaneously increasing the percentage of oxygen in the Earth's atmosphere and decreasing the percentage of carbon dioxide, plants paved the way for the rise of land animals."[226]

---

224 Bengtson et al., "Three-dimensional preservation," *PLOS*, March 2017.
225 Penn State, "Oxygen May Be Cause," *Science Daily*, Oct. 1999.
226 Heckman et al., "Molecular evidence for early," *Science*, 1129–1133.

So both Genesis and science agree that water appeared on earth, forming continents and oceans, and then plants appeared before animals—indeed, science suggests that plants played a role in preparing the atmosphere and making earth habitable for land animals.

DAY 4: LET THERE BE SEASONS

Genesis 1:14–19 And God said, "Let there be lights in the vault of the sky to separate the day from the night, and let them serve as signs to mark sacred times, and days and years, and let them be lights in the vault of the sky to give light on the earth." And it was so. God made two great lights—the greater light to govern the day and the lesser light to govern the night. He also made the stars. God set them in the vault of the sky to give light on the earth, to govern the day and the night, and to separate light from darkness. And God saw that it was good. And there was evening, and there was morning—the fourth day.

These verses point to the act of God in creating seasons, and in separating the day (light) from the night (darkness), so that humanity can enjoy their stay while on earth. For many of us, the existence of seasons and the occurrence of day and night might not be exciting, as we take it for granted. However, it took an incredible process to establish the seasons and our rhythms of life that go with them.

Our solar system is formed out of a spinning nebula that collapsed into a disk, all eight planets revolve around the center portion of the disk (our sun), rotating in the same direction because the original solar Nebula was rotating in that direction. All the planets also rotate about their axis in this same direction except Venus and Uranus. At present the earth is both rotating at 1000 miles/hour on its axis, completing one cycle every 24 hours, and also revolving around the sun at 67000 miles/hour. The earth is spinning (daily) on an axis which is tilted 23.5 degrees, compared with its orbit (yearly) around the sun. Meanwhile, the sun, with earth and the other planets, is rotating around the Milky Way galaxy at 490000 miles per hour.

it took an incredible process to establish the seasons and our rhythms of life that go with them

The Significance of Earth's Tilt

The Earth's two motions, namely rotation (spinning on its axis) and revolution (orbiting the sun), are the reasons for day and night and the changing seasons. "Earth has seasons because our planet's axis of rotation is tilted at an angle of 23.5 degrees relative to our orbital plane—the plane of Earth's orbit around the sun . . . The angle of tilt causes the Northern and Southern Hemisphere to trade places throughout the year in receiving the sun's light and warmth most directly."[227] However, we carry on with our daily life without thinking about all these happenings. If we just close our eyes and imagine, what if there were no seasons (spring, summer, autumn, and winter)?

If our earth were not tilted, there would not be any seasons, and the biosphere of the world as we know it would be entirely different. A recent article reports that, if there were no seasons, "humans would probably never have advanced past a state of living in small, scattered settlements, scrounging for survival and often dying of horrific insect-borne diseases."[228] This is because, without seasons, the farthest end of north and south latitudes would be chill all the time, while the equator would be much hotter than now, all the time. We would probably not have migratory birds or animals. The most important problem of our earth with no seasons would be that all of earth's population would have to live in a very small land surface, as most of the land would be covered with water due to raised sea level. Overall, the survival on earth would have been much harsher. Because of an elliptical orbit with a tilted axis, every part of the earth can receive sunlight at a regular interval, helping the sustenance of life on earth.

According to an article, "When a Mars-size object collided with Earth 4.5 billion years ago, it knocked off a chunk that would become the moon. It also tilted Earth sideways a bit, so that our planet now orbits the sun on a slant. Those were two huge changes. Now, throughout the year, the amount of sunlight striking the Northern and Southern hemispheres varies as they wobble back and forth—first the Southern Hemisphere leaning sunward, then the Northern. This

---

227   Conners, "Why Earth has 4 seasons," *EarthSky*, Sept. 2016.
228   Wolchover, "What If There," *Live Science*, March 2012.

cycle drives Earth's seasonal variation. It's a lucky thing, too. Without Earth's tilt, humanity would be in a sorry state."[229]

The Significance of the Moon

There is a theory that there was a collision of a Mars-size object with Earth around 4.5 billion years ago, knocking off a portion that would become the moon. This is known as the Giant Impact hypothesis, and it is the leading contender among many theories that have been put forward for explaining the formation of the moon. About 150 million years after the formation of our solar system, while many collisions were happening between planetary embryos, either one giant collision or "a succession of a variety of smaller collisions,"[230] would have given rise to the earth's tilt and the formation of the moon. From the earth's horizon, we can only see one side of the moon's hemisphere all the time as it is tidally locked in its revolution around the earth, that is, it takes 28 days to both go around the sun and to rotate once around its axis.

Without the moon, nights would have been much darker and one year would have contained more than 1000 days, with a day lasting somewhere between six and twelve hours. Also, the ocean tides would have risen much higher, resulting in huge waves. The gravity of the moon is responsible for driving the tides by pulling hard on the surface of the earth facing it. As the moon moves around the earth, the water covering the earth also moves along with it. Without the moon, the earth's oceans would be moved by the sun's gravity, with water rushing towards the side of the earth nearest the sun, creating huge waves.

"Around four billion years ago a cooling Earth already had an ocean but remained barren. The moon perhaps was half as distant as it is now, and as a result, the ocean tides were much more extreme. The oceans' tidal flow helps transport heat from the equator to the poles."[231] The moon is slowly leaving the earth's gravitational pull by a few inches per year. "The Moon is now about 384,000 km or 60 Earth

---

229 Wolchover, "What If there."
230 Rufu et al., "A Multiple-Impact," *Nature Geoscience*, 89–94.
231 Dorminey, "Without the Moon," *Scientific American*, April 2009.

radii away from Earth, which is about fifteen times further away than it was when it was first formed. The exact rate of the Moon's movement away from Earth has varied a lot over time. It depends both on the distance between the Earth and the Moon, and the exact shape of the Earth. Right now the moon is more than 238,000 miles from Earth, but when it formed, it was just 14,000 miles away."[232]

Though I knew that the moon is responsible for driving tides, I never realized until recently how important this positioning of the moon is. Even before I realized the full importance of the moon, it was always a point of fascination for me, ever since my childhood. My mother used to feed me by asking me to talk to the granny who is sitting inside the moon and stitching. In every culture, the moon has always been the subject of many myths and legends; associated with magic, folklore, superstitious beliefs and ancient rituals because the moon is a powerful force of nature. On a full moon day, when the sky is clear, no one can deny that just looking at the moon feels magical and amazing. As a kid, I tried many times to figure out this "granny" my mom told me about. I still can't make her out. But I see something else; I see the history of my planet engraved there—how a giant impact, a literal bang, has gifted this wonderful companion to me—to Earth. Somebody said that—*we live on a blue planet that circles a ball of fire next to a moon that moves the sea and you don't believe in miracles.*

> *God orchestrated this violent impact, giving rise to the earth's tilt, which is responsible for seasons, and to the moon, which governs the tides*

God orchestrated this violent impact, giving rise to the earth's tilt, which is responsible for seasons, and to the moon, which governs the tides. Later, He positioned the moon at the right place and, when God revealed this to the human author of Genesis, the author was thrilled, seeing the sun by day and moon and stars by night.

The Appearance of Stars

Did God make stars at this stage? The stars mentioned at day four might be stars that had formed earlier or stars that were forming right

---

232   Nevada's Space Site, "How Far Was The Moon."

then. God might even have revealed a "starburst" to the author of Genesis 1. A starburst region of space is an area with a high rate of star formation—a rate much higher than normally observed. A star forms in the interstellar space, when a dense region within the molecular clouds collapses in on itself under the influence of its gravity. A starburst can occur in various regions of space, even in entire galaxies.

Here are some details from a recent article on star formation— Observations by an international team of scientists, using the Very Large Array (VLA) in New Mexico, suggest that the central few light-years of the Milky Way has a population of about 200 massive, young and very bright stars, in tight orbits around Sagittarius A*, which are only a few million years old.[233] Scientists are wondering how these stars managed to form so close to a super-massive black hole. Even so, detecting these younger stars close to Sagittarius A* shows that there is an ongoing process of star formation.

The earlier "days" of the creation account are not governed by the sun, as though the sun (and earth's orbit) could establish the rhythm of God's work. Rather, the Genesis account makes it clear that it is God who created sun, moon, and stars and established their rhythms. The Genesis account also emphasizes that these "heavenly bodies," which many ancient peoples worshipped as divine, are only created things—simply "lights", not making an appearance at the beginning of the creation process, but only mentioned here, in the midst of all the other things God created.

Although science indicates that the creation of stars, planets, and moons began earlier, the igniting of new stars and the positioning of the moon were, nevertheless, ongoing processes. The seasons depend on the earth's orbit and spin, with its tilt as a vital factor. The moon's orbit also affects, most noticeably on tides. It is appropriate for sun, moon, and stars to be mentioned at this point because it is only after the plants begin to appear, the full significance of the seasons, which are governed by these celestial orbits, becomes apparent and can be appreciated.

---

233 Physics World, "Evidence for recent star," Mar. 2015.

DAY 5: LET WATER AND AIR TEEM WITH LIFE

Genesis 1:20–23 And God said, "Let the water teem with living creatures, and let birds fly above the earth across the vault of the sky." So God created the great creatures of the sea and every living thing with which the water teems and that moves about in it, according to their kinds, and every winged bird according to its kind. And God saw that it was good. God blessed them and said, "Be fruitful and increase in number and fill the water in the seas, and let the birds increase on the earth." And there was evening, and there was morning—the fifth day.

The above verses tell us that, after preparing everything for the sustenance of life, God was in this epoch in the process of creating various species that move in the water and that fly in the air. Science allows us to investigate the specific species and to estimate the dates of the eras when they appeared.

The end of the Paleozoic Era (352–252 million years ago) marked the greatest mass extinction in the history of life on planet Earth. "At the end of the Permian period, something killed some 90 percent of the planet's species. Less than 5 percent of the animal species in the seas survived. On land less than a third of the large animal species made it. Nearly all the trees died."[234] Geologists believe that climate changes due to the emerging supercontinent Pangaea were most likely the cause of the Permian extinction.

The Mesozoic era that began with the end of Permian extinction is divided up into the Triassic, Jurassic and Cretaceous periods. During the late Triassic period, almost all the Earth's land mass that formed the single supercontinent Pangaea had already started to rift and reptiles, that ruled the earth for over 150 million years, appeared. The fossil record shows that birds appeared during the Mesozoic Era. The fossil record has also indicated that a new lineage of fish called lobe-fins was present during the Paleozoic era. "Lobe-fins experienced considerable success during the Paleozoic Era. They exhibited a greater diversity than the ray-fins during the Devonian and Carboniferous Periods, and they were typically the top predators in many of the marine and freshwater habitats they occupied."[235]

---

234 Hoffman, "The Permian Extinction," *National Geographic.*
235 Devonian Times, "More About Lobe-fins."

Overall, the fossil record affirms the Bible's description that, once plants were established, creatures emerged in a wonderful variety of forms (although not all survive today), in the sea and then in the air. The fossil record gives us rather more evidence about the land creatures than it does about sea creatures (although these are dug up in places that are no longer covered by the sea) or about birds, whose light bones are rather rarely found as fossils.

DAY 6: LET THE LAND ABOUND WITH LIFE

Genesis 1:24–27 And God said, "Let the land produce living creatures according to their kinds: the livestock, the creatures that move along the ground, and the wild animals, each according to its kind." And it was so. God made the wild animals according to their kinds, the livestock according to their kinds, and all the creatures that move along the ground according to their kinds. And God saw that it was good. Then God said, "Let us make mankind in our image, in our likeness, so that they may rule over the fish in the sea and the birds in the sky, over the livestock and all the wild animals, and over all the creatures that move along the ground." So God created mankind in his own image, in the image of God he created them; male and female he created them.

Earlier, on "day 3", God spoke twice, and there were two acts of creation and now, on "day 6," again there are two: the creation of land animals and, then, separately, of humans. Within the fossil record, the most impressive remains are those of the dinosaurs, with their estimates ranging from about 1,000 to 10,000. Richard Owen coined the word "Dinosauria" from the Greek words "deinos" (fearfully great), and "sauros" (lizard), in 1841.

Origin of Animals

Genesis treats animals and humans as coming at the end of this long sequence. The fossil evidence supports this, but paleontologists are still exploring why animals have appeared so late in the record.[236]

> During the Mesozoic, or "Middle Life" Era, life diversified rapidly, and giant reptiles, dinosaurs, and other monstrous beasts roamed the Earth. The period, which spans from about 252 million years

---

236 Budd, *"The earliest fossil record,"* Philos Trans R Soc., 1425–1434.

ago to about 66 million years ago, was also known as the age of reptiles or the age of dinosaurs . . . The Mesozoic Era began roughly around the time of the end-Permian extinction, which wiped out 96 percent of marine life and 70 percent of all terrestrial species on the planet. Life slowly rebounded, eventually giving way to a flourishing diversity of animals, from massive lizards to monstrous dinosaurs. The Triassic Period, from 252 million to 200 million years ago, saw the rise of reptiles and the first dinosaurs, the Jurassic Period, from about 200 million to 145 million years ago, ushered in birds and mammals, and the Cretaceous Period, from 145 million to 66 million years ago is known for some of its iconic dinosaurs, such as Triceratops and Pteranodon.[237]

There were major extinctions at both the beginning and the end of the Triassic period. It began in the wake of the Permian–Triassic extinction event and ended with the Triassic/Jurassic extinction, a natural disaster that wiped out most, perhaps even 90 percent, of the planet's species. Volcanoes, a meteorite, global cooling, or some combination of these, have all been suggested as possible reasons for these extinctions. Almost half of the species known to have been living on Earth at that time became extinct, allowing dinosaurs to become dominant during the Jurassic period.

It was during the Jurassic period that the Pangaean supercontinent broke into two parts–Gondwana and Laurasia. "Gondwana was an ancient supercontinent that broke up about 180 million years ago. The continent eventually split into landmasses we recognize today— Africa, South America, Australia, Antarctica, the Indian subcontinent and the Arabian Peninsula . . . Gondwana was half of the Pangaea supercontinent, along with a northern supercontinent known as Laurasia."[238]

The Cretaceous period is the longest period of the Mesozoic Era when the earth began to assume its modern form. At the end of the Cretaceous period, geologists believe that an asteroid (or comet) impact on the Earth along with a volcanic activity triggered shockwaves and massive tsunamis sending a large cloud of hot rock and dust into the

---

237 Ghose, "Mesozoic Era: Age," *Live Science*, Jan. 2015.
238 Pappas, "What is Gondwana?" *Live Science*, June 2013.

atmosphere, blocking part of the sun's ray for months or years, causing most of this vegetation to die out.[239] Most of the herbivores died due to this loss of vegetation eventually causing the death of the carnivores which ate the herbivores. This heat blast along with a rain of hot dust would have caused forest fires killing animals too large to find shelter, while smaller animals would have got shelter underground, in caves or tree trunks. This Cretaceous-Paleogene extinction marked the end of the dinosaurs.

The current era is the Cenozoic era, from 65 million years ago to the present. The smaller mammals, that survived the extinction event that killed all the dinosaurs, became dominant during this era. Also, the climate changed drastically; it became much cooler and drier in a relatively short period. Our earth also experienced an ice age event with most temperate parts of the Earth covered in glaciers. Today there are about 5,000 species overall, and they are characterized by their hair or fur, the milk with which they suckle their young, and their warm-blooded metabolisms. The Cenozoic Era will go on until there is another mass extinction.

Origin of Humans

God created a wide range of different species of animals, some of which are alive today while others have, at various stages, become extinct; and then He created human beings, the pinnacle of His creation—who came into existence during the Cenozoic Era. The idea that humans existed only for the last 6,000 years was "arrived at by James Ussher, a 17th century Irish Archbishop who counted up estimates of the ages of Abraham's family listed in the Old Testament and calculated that the creation began (on the Julian calendar) on Saturday, October 22, 4004 BC, at 6 pm."[240] However, William Henry Green argues that "the genealogies in Genesis 5 and 11 were not intended to be used and cannot properly be used for the construction of a chronology . . . the Scriptures furnish no data for a chronological computation before the life of Abraham; and that the Mosaic records do not fix and were not intended to fix the precise date either of the flood or of

---

239 Sokol, "Volcanoes plus asteroid," *New Scientist*, Oct. 2015.
240 Huffingtonpost, "Can We Measure the Earth's Age."

the creation of the world."²⁴¹ In support of this, few theologians say that "the genealogies have nothing to do with the actual length of the overall period and therefore it is unnecessary to press them into a rigid chronological system."²⁴² Green's suggestion, accepted by many Christians ever since, was that the Bible does not specify the age of the universe, which leaves Christians free to consider scientific evidence and draw conclusions from it.

> the Bible does not specify the age of the universe, which leaves Christians free to consider scientific evidence and draw conclusions from it

Ann Gauger, a senior research scientist in molecular genetics considers a controversy: whether the origin of the first human beings could potentially go back to a single pair, or whether a "bottleneck" of 10,000 or more individuals seems required by population genetics. The argument from population genetics has been that there is too much genetic diversity to pass through a bottleneck of two individuals, as would be the case for Adam and Eve. But Ann Gauger suggests that this is not true and she is convinced about the possibility of a literal Adam and Eve. She believes that the possibility of all human life coming from two parents is much greater than all life descending from a primordial cell. She concludes, "I chose to look at the HLA-DRB1 story because it seemed to provide the strongest case from population genetics against two first parents. If it were true that we share thirty-two separate lineages of HLA-DRB1 with chimps, it would indeed cause difficulties for an original couple. But . . . the data indicate that it is possible for us to have come from just two first parents . . . one thing is clear right now: Adam and Eve have not been disproven by science, and those who claim otherwise are misrepresenting the scientific evidence."²⁴³

Science points to the fact that the Edenic pair is not an allegory or a myth and that the human race did arise from a single set of parents. The Bible tells us that human beings are made in God's very image and that their main purpose is to have a continual fellowship

---

241 Green, "Primeval Chronology," *Bibliotheca Sacra* 47, 286, 303.
242 Whitcomb Jr., and Morris, *The Genesis Flood*, 474–489.
243 Gauger, et al., *Science and Human Origins*, 120,121.

with Him, both now and forevermore. Science has unearthed many facts about human beings and has found that we are distinct in many aspects from other species. We have some similarities to animals, especially to chimpanzees, both in our genes and in our social behavior, but there are also huge differences between humans and any other species. When getting food, some animals, including chimpanzees can use rocks as tools; but humans can make complex tools and to control their food supply by growing, harvesting, storing, processing and transporting it globally. Many species are social and communicate, but only humans can communicate and record complex ideas. Only humans devote time and effort to art that has no survival value or functional purpose. Finally, human beings alone possess conscience and morality, with a desire to worship and a sense of justice, experiencing shame and guilt.

> *Science points to the fact that the Edenic pair is not an allegory or a myth and that the human race did arise from a single set of parents*

So, although Genesis does not take time attempting to describe creatures, such as dinosaurs, which would be totally unfamiliar to the people it was originally written for, the sequence of its account does, nevertheless, fit with what the scientific evidence indicates—that creatures in the sea were followed by all kinds of creatures on land, with humans only emerging comparatively recently. Genesis treats humans as created separately, after animals, and this distinction can be seen by comparing human and animal behavior—socially, in having and communicating ideas, in morality and self-awareness.

That is the conclusion of the process of creation, but it is not quite the end of the "days" of this fascinating sequence.

DAY 7: LET THERE BE REST

Genesis 2:2–3: By the seventh day God had finished the work he had been doing; so on the seventh day he rested from all his work. Then God blessed the seventh day and made it holy, because on it he rested from all the work of creating that he had done.

Many are intrigued by Genesis 2:2–3, as it says that God completed His work and He rested on the seventh day. Did God need a day off because He was exhausted? No. Scripture is very clear that God does not need rest; He can never be tired since He is omnipotent. Has God completed all His works? We can say both Yes and No. It is yes because He completed His work towards creating His special creation, both male and female, made in His image, and He also completed His work by giving them a beautiful habitat to live. However, the answer is also no, because God is still active in His work of redemption and providence.

The Hebrew word *Shabbat* translated "rested" in Genesis 2:2 does not necessarily include the idea of being tired; it means "to cease or stop."When it says He "stopped," it means the revelations given to Moses ended. Now Moses, after six days of intense visions and dreams, could rest in God. In other words, we can say that God instituted rest as a necessary component for human beings and as a part of his weekly routine. In my understanding, an account of this revelation and rest is given in Exodus 24:12–18. Moses climbed the mountain as commanded, to receive the tablets of stone and the law and commandments. The glory of God rested on Mount Sinai (God rested), and for six days a cloud obscured the mountain. God could have shown the Creation account to Moses consecutively on all those six days. On the seventh day, from the cloud, God summoned Moses; and Moses entered the cloud and was there forty days, and forty nights (Moses rested). Moses may well have enjoyed both physical and spiritual rest in God's presence after his intense spiritual experiences for six full days.

Both physical and spiritual rest is needed, as we are bombarded with activities every waking moment and our hearts are often restless in yearning for something that can satisfy us. On June 3, 1965, Time Magazine proclaimed, in its headline about the spacewalk of Edward White.[244] Walking in space must have been just plain cool because, when he was ordered to get back into his capsule, he protested like a scolded kid. "I'm doing great," he said. "It's fun. I'm not coming in." When, after 20 minutes of space gymnastics, he finally did agree

---

244 http://time.com/3898131/edward-white-walk-space/

to squeeze himself back into his Gemini 4 ship, he said to Command Pilot James Alton McDivitt: "It's the saddest day of my life," as he still had not had enough of spacewalking. Like Ed White, most of the time, any amount of adventure or romance does not satisfy us; we are restless inside our hearts. Augustine says that our hearts are restless until it finds its rest in God. Jesus Christ says, "Come to me, all you who are weary and burdened, and I will give you rest." (Matthew 11:28–30). Only after surrendering one's life to Christ, and developing a complete trust in God, can a person enter into God's rest. Entering into complete rest in God is a beautiful process. It is a lifetime journey. It is living in His Presence moment by moment. It is not a one-time affair. Trust precedes rest, the most incredible thing a person can achieve in his life. Trusting in God helps a person to enjoy the security of God's unchanging love and to be sure of eternal inheritance; the assurance of salvation a certainty which Christ alone provides (1 John 5:13). This knowledge, that life has a great purpose ordained by God and has an eternal destiny, is life-changing and powerful.

> *Entering into complete rest in God is a beautiful process. It is a lifetime journey. It is living in His Presence moment by moment*

After witnessing a glimpse of the act of God creating the whole universe with His awesome power and might, Moses would certainly have enjoyed this rest in God. Achieving this rest is the greatest thing anyone can achieve during their journey of life. Genesis includes no hint of an eighth day. It seems to treat God's rest as continuing now, and people are invited to share in that in two ways—by having one day of rest (a "sabbath") every week, and by taking up God's invitation to join Him in His eternal rest.[245]

CONCLUSION

The journey of the story of the cosmos is now completed and this journey shows how science and Scripture can support and complement each other while the story unfolds. Genesis gives a structured, step-by-step account of the origin of the universe and the origin of life,

---

245 Hebrews 4:11; Revelation 14:13

arranged so that the process of creation is described over six "days." The "revelation model" is used to interpret the biblical account as describing, not six 24-hour periods, but six epochs. The creative processes described in Genesis have been related to the scientific evidence available about this sequence. Science provides more detail (e.g., evidence of extinct species, especially dinosaurs), but Genesis provides more connections with questions of meaning and purpose. This table gives the comparison of the sequence as told by science and the Bible in a nutshell.

Sl No	Creation Sequence as per Bible	Epochs of the Cosmos as per Science (Approx.)**	The Story of the Cosmos
1	Let there be light *Genesis 1:1–5*	13.7 billion years ago	The Big Bang and the expansion of the universe
		3 to 20 minutes after the Big Bang	The 'Whole Universe is a Star' condition
		380000 years after the Big Bang	The Era of Recombination: The first light photons were released
			The Cosmic Dark Age (150 million years)
			The Epoch of Re-ionization: The birth of first stars – Population III stars
			The Cosmic Renaissance: The birth of Population II and I stars
2	Let there be a habitable planet *Genesis 1:6–8*	4.6 billion years ago	Positioning our solar system in the inner rim of the Orion arm of Milky Way Galaxy
			Positioning our planet inside our solar system
			Creation of protective layers (Heliosphere and the Earth's Atmosphere)

#	Day	Time	Events
3	Let there be life Genesis 1:9–13	3.8 billion years ago	Origin of oceans and origin of water on Earth—comets/asteroids Origin of continents
		2.6 billion years ago	Emergence of first life and plants.
4	Let there be seasons Genesis 1:14–19	3.5 to 4.5 billion years ago	Formation of Moon by the Giant Impact Positioning of Moon at the right orbit Earth's tilt—23.5 degrees relative to our orbital plane Star burst
5	Let water and air teem with life Genesis 1:20–23	352 to 252 million years ago	Appearance of birds and sea creatures—Paleozoic Era; The Permian-Triassic Extinction
6	**Let the land abound with life** **Genesis 1:24–31**	252 million years ago to about 66 million years ago	Appearance of land animals— The Mesozoic Era (Triassic, Jurassic and Cretaceous periods); Triassic-Jurassic Extinction; Cretaceous-Paleogene Extinction Appearance of man (male and female)—
		No scientific consensus	Cenozoic Era
7	Let there be rest Genesis 2:2–3; Hebrews 4:1–11		Ongoing

**   The above table gives the epochs as per the current scientific consensus about the ages of the universe, the solar system and life on earth. In time, this consensus may change; but the sequence given in the other columns will not vary.

PART 5
EPILOGUE

CHAPTER 9

A SHOWDOWN WITH TRUTH

If the doors of perception were cleansed everything would appear to man as it is—Infinite.

William Blake

FINDING TRUTH

Is it possible to find the truth about our origins—where we come from? This has been the heart of this book. Finding the right answer to this ultimate question would help in finding the answer to the other ultimate questions: What is the meaning of life? Where do we end up when we die? What is the future of this material world that we have investigated—is it heat death or something to look forward to? Throughout the ages, people have been searching for answers to such questions and wondering whether they have found the truth.

But what is "truth"? This profound and eternally significant question was posed by Pilate, during the interrogation of Jesus. Pilate asked, "What is truth?" But, for him, this was only a rhetorical question, a cynical response to Jesus' statement: "I have come into the world, to testify to the truth." This historic encounter raises the perennial question of the very nature of truth itself. The vast majority of philosophers and theologians throughout history regard truth as what corresponds to reality, without being distorted by a person's beliefs and opinions. Our search for truth should not be a mere rhetorical question, which expects no reply; and it should not be a mere re-statement of our prejudices. Even scientists, who strive for objectivity in their analysis, can be influenced by their culture, experience, and beliefs. So, while we search for truth, we should be aware of the danger of our subjective perceptions not correlating with the objective reality.

I set out on this interesting journey—to find the missing pieces of the puzzle behind the question of the origin of the cosmos and life on earth. I aimed to use an approach that would not be influenced by my perceptions and opinions, and so I gathered evidence using a cumulative approach. I focused on the following key questions: How did the Big Bang come into effect—and so what happened before the Big Bang? What explains the order, regularity, and fine-tuning of physical laws and constants which allows life to exist on this planet earth? How did the first life come into existence? How all the various life forms with enormous complexities emerge?

I searched for evidence from the latest research in the various disciplines of science, namely archaeology, physics, biology, geology, genetics, and cosmology. To discover the scientific evidence about the origins of the cosmos and of life, I journeyed through the Big Bang, through the birth and death of stars, through the formation of galaxies and our sun, through the origin of water on our planet, and the beginnings of living organisms—single-celled organisms, multi-cellular organisms, fish, birds, animals and human beings. Learning about the major mass extinctions, the ice ages, the life and fall of many creatures like dinosaurs took me on an interesting path. The many unanswered questions fascinated me, including known unknowns such as the conditions during the universe's beginning, fine-tuning and the origin of the first life. To tell the story of the cosmos and life scientifically, this book has had to navigate through the demanding rigors of science, but I have tried to make this as clear as possible.

For explaining the cause of the Big Bang, none of the scientific theories put forward – such as the quantum gravity model, the no-boundary proposal, the oscillating universe model and the universe tunneling out of nothing—are up to the task of giving a credible explanation. Most of them involve unsubstantiated notions, like the spontaneous fluctuations in a quantum vacuum, the concept of dark matter and dark energy and so on. The complete explanation of reality through science seems impossible since all endeavors for developing the theory of everything by uniting the two great theories of the 20th

century—General Relativity and the Quantum Physics—have not yet achieved success. Even if such a theory were available, no experimental set up on earth is available to test the predictions of such a theory. The question of how the universe works now, as well as the question of what came before the Big Bang remains mysterious and unanswered.

The next question—the explanation behind the order, regularity, and fine-tuning of physical laws and constants to allow life to exist—suggests the existence of an intelligent designer. Our universe cannot be a chance universe as put forth by the multiverse hypothesis. The multiverse hypothesis is highly speculative because it is not possible to demonstrate whether or not other universes exist, even if technology improves; and, even if they do exist, there would also have to be some mechanism by which all those universes came into existence. Richard Swinburne says "To postulate a trillion trillion other universes, rather than one God in order to explain the orderliness of our universe, seems the height of irrationality."[246] There is a strong reason to wonder whether there is any other explanation, any outside phenomenon beyond the physical reality, for the beginning conditions of the universe and the fine-tuning of the Cosmos.

For the final question—can science explain the origin of life and the complexities of various life forms, the simple answer is—the origin of life with its complexities is shrouded in mystery. The theories put forward, like the primordial soup theory, hydrothermal vent theory, and the Panspermia theory, to explain the origin of life from non-life (abiogenesis)—are speculative theories. Science is unable to give a credible explanation for the rise of life spontaneously by random chance.

Richard Dawkins says that "Natural selection, the blind, unconscious, automatic process which Darwin discovered, and which we now know is the explanation for the existence and apparently purposeful form of all life, has no purpose in mind. It has no mind and no mind's eye. It does not plan for the future. It has no vision, no foresight, no sight at all."[247] However, the blind process discovered by

---

246   Swinburne, *Is There a God?*, 68.
247   Dawkins. *The Blind Watchmaker*, 5.

Darwin has not proved itself, as many expected, because it has ended up in many contradictions. Fred Hoyle and Chandra Wickramasinghe after their investigation said that "there are so many flaws in Darwinism that one can wonder why it swept so completely through the scientific world, and why it is still endemic today."[248] The construction of the evolutionary tree for arriving at human lineage is highly ambiguous as interpretations vary among Darwinian scientists, e.g., interpretation of fossils has been changing with time, and molecular clock gives different data with different estimates. Also, the construction of the evolutionary tree by tracing the lineages would need many, complicated steps. The fish-to-amphibian evolution, the dinosaur-to-bird evolution, land mammal-to-whale evolution, apes-to-human evolution and so on, lacks not only much support both in the fossil record and molecular data, but are also complicated. Having considered the evidence and the theories, the idea of universal descent remains far from convincing.

The major conclusion of my findings, concerning the origin of the universe and the origin of life, is this—Although it sometimes feels as though we now know a lot about the universe, actually, there is far more than we do not know. The cumulative epistemological findings, made it clear, that a purely scientific explanation for the origin of cosmos and life were far from convincing. The scientific advances have only pointed out that there are considerable gaps in our scientific understanding. Our ignorance on some of the subjects, like the chemical evolution of the universe and the biological evolution of life, is vast. There are unbridged gaps in our understanding and in the existing theories, some of which may never be bridged. We know what we perceive, but there is every reason to believe that reality has a depth to it that extends beyond our understanding.

> *Our ignorance on some of the subjects, like the chemical evolution of the universe and the biological evolution of life, is vast*

I found that, though science leads to great achievements, it cannot explain everything, because science is not the only source of reliable information or a valid form of knowledge. In the human quest for truth, science is only one approach.

---

248   Hoyle and Wickramasinghe, *Evolution from Space*, 133.

Other sources may also be sought to grasp the complete truth. So in my search for the truth, in due course, I used not only science, but also philosophy and religion to probe deeply into the past, to search for the story of the Cosmos and of life, the story of how the universe came into existence and how life with intelligent and moral beings began. Among the various ideas put forward in trying to understand our origins, the idea of an intelligent designer provides the most straightforward explanation.

THREE WAYS OF TESTING WORLDVIEWS

I believe that the findings that I have presented here, based on reliable and compelling evidence have a reasonably clear and unbiased perspective and will help you to discern truth from falsehood. But how can you work out what to believe?

We must bear in mind that our preconceptions dramatically influence the way we perceive the world. We see the world not as it is, but only through the lenses of our perceptions; through the glasses of our worldview. Worldview is one's perception of actual or potential realities. James W. Sire defines it as: "A worldview is a commitment, a fundamental orientation of the heart, that can be expressed as a story or in a set of presuppositions (assumptions which may be true, partially true or entirely false) which we hold (consciously or subconsciously, consistently or inconsistently) about the basic constitution of reality, and that provides the foundation on which we live and move and have our being."[249] A worldview may be subconscious or clearly thought out beliefs and values (i.e.); it may be sensible or absurd, but it is so profound that it governs the way we think and guides our decisions. Since it colors the way we see almost everything around us, different people can view the same reality in diverse ways.

Naturalism is a worldview that sees the universe as governed by natural laws and random chance and not by any supernatural being or force. There are many religions, which lead to a variety of worldviews. Two significant religious worldviews can be summarized as theism and pantheism. Theism regards the universe as having been created

---

249   Sire, *The Universe Next Door*, 17.

by a personal God. Pantheism is the view that the natural universe is divine and there is no separation between creation and god. The god of pantheism is an impersonal god—god is all, and all is god (pan = all and theism = god). According to pantheism, the intelligence that brought the universe into existence is part of the universe. Because of the impersonal beginning of the universe, this view is left out of the discussion, and the main argument presented in this book is: whether the universe and life can originate with or without an eternal supernatural being.

How is it possible to weigh up these different worldviews? There are three approaches, each of which seems to be credible and widely acceptable. So I have subjected the naturalistic worldview, with no supernatural being, and the theistic worldview, that believes in a supernatural Creator, to these three tests:

a. Occam's Razor

b. Falsifiability Criteria

c. The 3-4-5 Grid Method

Occam's Razor

William of Ockham, of the 14th century, said that, if there are two explanations for a single problem, the simpler one is usually the better, or one should choose the explanation that makes the fewest assumptions. Ernst Mach formulated a stronger version of Occam's razor and called it the Principle of Economy: "Scientists must use the simplest means of arriving at their results and exclude everything not perceived by the senses."[250] Many scientists use this concept to shave away any metaphysical concepts. Stephen Hawking writes: "We could still imagine that there is a set of laws that determines events completely for some supernatural being, who could observe the present state of the universe without disturbing it. However, such models of the universe are not of much interest to us mortals. It seems better to employ the principle known as Occam's Razor and cut out all the features of the theory that cannot be observed."[251] However, this view is self-contradictory because

---

250 Becher, "The Philosophical Views," *The Philosophical Review*, 535–562.
251 Sober, *Ockam's Razor: A User's Manual*, 4.

scientific theories put forward to explain the beginning condition of the universe and the fine-tuning argument, such as String theory and M-theory, are no simpler, but depend on the idea of dimensions, or even entire universes, which "cannot be observed."

A purely naturalistic approach does not help in explaining the mysteries of the origin of the universe. This approach not only fails to explain the mysteries, but they seem to be complicated and introduce a variety of seemingly unanswerable questions, leaving us with even more difficult questions, thus failing the test of Occam's Razor. The same razor that some suggest should shave out any metaphysical entity should, all the more, shave out the unobservable speculations involved in naturalistic explanations.

Criterion of Falsifiability

"Falsifiability" is one of the guidelines for distinguishing a scientific theory, from a non-scientific one. Falsification of a theory would show that it is not trustworthy. So falsifiability requires that there should be some method or experiment that could "falsify" the theory– a test which, if the theory is not true, would prove that it is not true. In the seminal work, Popper states that "the criterion of the scientific status of a theory is its falsifiability, or refutability, or testability."[252]

Roger Penrose, the mathematician who designed the famously "impossible" depictions like the "Ascending and Descending" staircase and the "Waterfall" and developed "Penrose tilings" said while speaking for the faith debate program "Unbelievable?" that Stephen Hawking's claims, that the M-theory can explain everything and has made God redundant as the first cause, are misleading.[253] Penrose called M-theory "hardly science" and "not even a theory," but only a collection of hopes and ideas, very far from any testability. Penrose, who subscribes to atheism, had this to say about the "multiverse" hypothesis that was posited as an explanation for the incredible fine-tuning of our universe: "It's overused, and this is a place where it is overused. It's an excuse for not having a good theory." Two cosmologists from John Hopkins University in Baltimore, Maryland, have written about a "worrying

---

252 Popper, *The Logic of Scientific Discovery*, 46.
253 Premier Christian Radio, "Unbelievable," Sept. 2010.

turn" in theoretical physics—"faced with difficulties in applying fundamental theories to the observed universe, some researchers have called for a change in how theoretical physics is done, arguing that, if a theory is sufficiently elegant and explanatory, it need not be tested experimentally. Those cosmologists have cited String theory because the infinitesimally thin strings are not detectable using today's technology and also, M-theory because the many universes postulated cannot be observed experimentally. But this breaks with centuries of philosophical tradition which has defined scientific knowledge as empirical."[254] Thus, most of the theories that have been given for the explanation of the beginning condition and the fine-tuning, such as String theory and M-theory, cannot be even called scientific theories because they are not falsifiable. This has become a source of great frustration for naturalistic scientists.[255]

> *most of the theories given for the explanation of the beginning condition and the fine-tuning ..... cannot be called scientific theories because they are not falsifiable*

Similarly, the existence of a supernatural being is also non-falsifiable. Questions about the existence of the Supernatural extend beyond the realm of nature and, therefore, of science. Science has limitations in drawing inferences regarding the existence of an invisible supernatural being—it can neither affirm nor deny His existence. This is because God exists outside the natural world and there are no experiments or measurements to prove His existence. C.S. Lewis states, "if there was a controlling power outside the universe, it could not show itself to us as one of the facts inside the universe—no more than the architect of a house could actually be a wall or staircase or fireplace in that house."[256]

> In investigating our origins, how should we cope with having limited evidence and ideas that lack falsifiability? Arguably, all those who take a view have had to make some leap of faith. The

---

254 Castelvecchi, "Feuding physicists," *Nature*, 446–447.
255 Castelvecchi, *"Is String Theory,"* Nature, Dec. 2015.
256 Lewis, *Mere Christianity*, Book 1, chapter 4, 20.

famous evolutionist Robert Jastrow said:

> Perhaps the appearance of life on the earth is a miracle. Scientists are reluctant to accept that view, but their choices are limited; either life was created on the earth by the will of a being outside the grasp of scientific understanding, or it evolved on our planet spontaneously, through chemical reactions occurring in nonliving matter lying on the surface of the planet. The first theory places the question of the origin of life beyond the reach of scientific inquiry. It is a statement of faith in the power of a Supreme Being not subject to the laws of science. The second theory is also an act of faith. The act of faith consists in assuming that the scientific view of the origin of life is correct, without having concrete evidence to support that belief.[257]

Both scientific and faith-based explanations of our origins are open to the charge of being speculative and lacking falsifiability. But, in the case of the biblical worldview, there may be another, relevant "cosmic event." Intriguingly, as well as these two cosmic events of the origin of the universe and the origin of life, the biblical worldview points to the resurrection of Jesus as the third event of cosmic significance—one which is open to investigation.

> *the biblical worldview points to the resurrection of Jesus as the third event of cosmic significance—one which is open to investigation*

The 3-4-5 Grid Method of Analyzing Worldviews

The 3-4-5 grid method, offered by author and speaker Ravi Zacharias, is about answering the four ultimate questions (origin, meaning, morality, and destiny) in such a way that the answers correspond to truth (by either empirical or logical reasoning), and they must be coherent (one part of the answer must not contradict another part).[258] In this approach, to check the credibility of any particular, proposed worldview, we can subject it to three tests:

a. Logical consistency—can it pass the test of reason or logic?

---

257 Jastrow, *Until the Sun Dies*, 62–63.
258 Zacharias, "Think again," *Slice of Infinity*, Aug. 2014.

b. Empirical adequacy—does it have verifiable observations?

c. Experiential relevance—does it work in real life?

A purely scientific approach certainly aims at being empirically adequate and logically consistent, but these break down when it comes to explaining the mysteries behind the origin of the universe and life. It has been brought out in this book, that there is no logical explanation for the beginning condition and the fine-tuning of the universe; nor for the beginning of life with its complexities. The theories put forward to explain these mysteries are not empirically adequate. For example, many of the tenets of neo-Darwinism have been falsified by many experiments and observations. Cornelius Hunter, of the Center for Science and Culture, has given twenty-two falsified aspects of Darwin's theory of evolution.[259] Some of the falsifying observations are—many species do not have sufficient genetic variation to respond to selection; many species seem to have appeared whole and fully formed, as they are today; complex organisms did not develop from simple life structures, because many organisms are, at the gene level, almost as complex as humans; and the fossil record shows clear evidence for leaps in life forms with no intermediate fossils.

> *The pictorial icon of evolution (the ascent of man)—a drawing depicting a steady progression from a crouching ape to a modern human—is not supported by the evidence*

The pictorial icon of evolution (the ascent of man)—a drawing depicting a steady progression from a crouching ape to a modern human—is not supported by the evidence. It oversimplifies a complexity that is beyond explanation.

Another challenge for neo-Darwinism comes in the accommodation of consciousness. It is not clear that such random processes can provide an explanation that is empirically adequate, to account for human consciousness and how it arose. If brains were shaped only for survival, and our cognitive capacities were merely evolved dispositions, then how can the brain, with its cognitive capacity, distinguish right from

---

259   Hunter, "Darwin's Predictions," 2015.

wrong? How can the brain decide whether naturalism or theism is true, based on the evidence, if the thought processes are based on just natural, random processes, including the firing of neurons?

On the other hand, Richard Dawkins suggests that, now that we have understood genes, the human race is a totally unique lifeform, in the whole of history. This is because we are now able to understand that we are the product of our "selfish genes." Our unique consciousness means that we are the only species that, rather than blindly supporting the survival of our genes, can decide for ourselves how to respond to our understanding of genes. So, even a scientist who believes in unguided, naturalistic evolution ends up noticing that our consciousness is unique and raises new questions.

According to Francis Crick, human beings are merely the accidental result of a mindless process of winning mutations and the mind is nothing more than the brain. He writes, "You, your joys and sorrows, your memories and your ambitions, your sense of personal identity and free will, are in fact no more than the behavior of a vast assembly of nerve cells and their associated molecules."[260] However, John Lennox, professor of mathematics at the University of Oxford, writes that if, "the brain is the end product of a mindless, unguided process . . . and you trust it? If you knew that your computer was the end product of a mindless, unguided process, you wouldn't trust it for a moment, would you? And yet to do your science, you trust something that you believe has come to be without any mind behind it whatsoever . . . Naturalism is fatally flawed. It undermines the foundations of the very rationality that is needed to construct or understand or believe in any kind of argument."[261]

Our minds are not governed just by the rules of science. Science depends on the precondition of intelligibility, but naturalism has no way to demonstrate that intelligence is anything more than a survival mechanism. Keith Ward says, "We are not just information processing systems. We are also conscious appreciators of the meaning

---

260   Crick, *The Astonishing Hypothesis*, 3.
261   Lennox, "Intellectually Fulfilling," Vol. 1, 2016.

of information, and creative initiators of new processes of thought."[262] Philosophers and scientists subscribing to neo-Darwinian evolution have been engaged in sharp disagreements for many decades over the issue of the formation of intelligence in the human mind and the mystery of consciousness, and this continues till now when neuroscience is incredibly sophisticated. What makes us humans? This question is the central mystery of human life.

The Bible does address this question. As the last chapter shows, the creation sequence given in Genesis 1 correlates with the sequence suggested by modern science. It also goes further and provides a logically consistent explanation of reality by answering the mysteries and questions beyond humanity, while beautifully explaining the cosmic story in a unified and comprehensible way. As science and Bible together provide answers to the question of origin, missing pieces of the puzzle have fitted together—from the small-scale to the large-scale grandeur of the universe and from microorganisms to human beings. Scientific evidence suggests that the universe had a beginning and also that it began as a highly ordered system, which fits with the second law of thermodynamics (that a system always becomes less ordered over time) and is also exactly consistent with the biblical worldview. So the Christian worldview, when it takes the scientific evidence seriously, is logically consistent and empirically adequate.

By providing a credible and compelling answer to that ultimate question of origin, the other three ultimate questions (meaning, morality, and destiny) can easily be answered. The Christian worldview is based on moral absolutes revealed by a transcendent God, regardless of people's subjective views. Since the foundation for ethics depends on the very character of God, as revealed in the Scripture, there is a solid base for determining right from wrong in a Christian worldview. On the other hand, if the answer to the first question of origin is random chance from a naturalistic worldview, then, the universe becomes purposeless, as human life in its cosmic setting is purposeless, with no meaning. Since naturalism suggests that the human race that emerged through blind processes will eventually evolve into something else, or die out, then life can also be extinguished blindly with no meaning

---

262 Ward, Why there almost certainly, 2008.

to our existence, and no hope of finding our loved ones who have departed from us. A purely naturalistic approach cannot explain the "why" questions of life—why we appreciate the beauty and why we love and why should be right and so on. Science cannot explain morality, beauty, art, and love because science does not accept many sensory experiences as observational data. Naturalistic worldviews ignore or lack the tools to address the question of morality because starting from randomness tends to lead to moral relativism, denying any authoritative moral code. When it comes to experiential relevance, a purely scientific worldview leaves significant gaps over questions of meaning and purpose, whereas the biblical worldview addresses those questions quite directly.

Conclusions from the Three Tests

A purely naturalistic worldview explains much around us but, when it comes to questions of our origins, it ends up with speculations which fail the tests of Occam's razor and falsifiability. On this basis, the idea of a supernatural Creator is no less credible. The Christian worldview is not only logically consistent and empirically adequate but also, unlike the naturalistic worldview, experientially relevant, with the power to transform a human being into a better individual.

> *The Christian worldview is not only logically consistent and empirically adequate but also, unlike the naturalistic worldview, experientially relevant.*

If the beginning of the universe had a cause, that cause had to be beyond time and had to exist independently of our universe. Also, if the laws of logic are absolute, they imply a logical being that brought it into existence. The profile of the biblical God fits properly with this logical being, who is the cause of the universe. For example, in John's gospel, when John refers to Jesus as the eternal logic (*logos*) of God, he also indicates that the Universe is mathematical in its foundation. Logic and mathematics are parallel disciplines. Galileo said that the universe is written in the language of mathematics. Both mathematics and logic must reflect the nature of the one who created the Universe. God's world has beautiful connections to elegant mathematical structures, and this mathematical beauty inherent in the universe compels us to acknowledge the wisdom and design behind it.

A purely naturalistic worldview can give the impression of plausibility by claiming to be based only on evidence. This may be true of many aspects of modern science but, when it comes to these questions of origins, this approach ends up invoking various speculations which, on inspection, are far from plausible—such as a multitude of unseen dimensions or even universes. When it comes to the question, "What is the meaning and purpose of life?" it is not clear that this approach has an answer to offer, because the very concept of "purpose" is excluded from our origins.

On the other hand, the Bible is shown to be a historically accurate document by thousands of archaeological discoveries and secular historical records. Many of the archaeological finds, the writings of historians and critical analysis of the gospel writers provide strong evidence for the empirical adequacy of the Christian worldview.[263] Though written over a period of about 1500 years, by over forty different authors, it does not contradict itself from the beginning to the end. The Bible also offers a particular explanation of the meaning and purpose of the universe and of life, which it claims is experientially relevant—relevant to the life and experiences of each one of us.

> The same God who has left His signature in the Bible has also left His signature in nature

The same God who has left His signature in the Bible has also left His signature in nature. That is why natural beauty has a profound effect upon our senses, and if only the perception of our eyes can be well adjusted, we will stand in awe and wonder with an astonished gaze at the transcendent glories of spiritual realities. Also, it is possible to look at this universe, with all the scientific evidence about its nature and origins, and conclude that the idea of a supernatural Creator is utterly foolish, responding, like Bertrand Russell, "You didn't give us

---

263   Greenleaf, *An Examination of Testimony*, 29.

enough evidence." But, if the Bible is, as Christians claim, the Word of God, then we can expect it also to provide evidence, stretching the intellect and informing the human mind by soaring to the heights of knowledge and plumbing deep into the recesses of the heart.

SEEING THROUGH CLEAR GLASSES

Many scientists who are committed to materialism do not want anything supernatural to be part of the story, no matter whether the evidence supports it or not. "Evolution is a theory universally accepted, not because it can be proved by logical, coherent evidence to be true, but because the only alternative, special creation, is clearly incredible."[264] Dr. Scott Todd writes, "Even if all the data point to an intelligent designer, such a hypothesis is excluded from science because it is not naturalistic."[265] Though the evidence points to the supernatural, many naturalistic scientists don't want to accept it due to their worldview. Professor Richard Lewontin, a leading evolutionary geneticist, writes, "We take the side of [evolutionary] science despite the patent absurdity of some of its constructs . . . in spite of the tolerance of the scientific community for unsubstantiated just-so stories because we have a prior commitment, a commitment to materialism . . . Moreover, that materialism is absolute, for we cannot allow a Divine Foot in the door."[266] This startling statement reveals that many scientists like him who are committed to materialism may be more interested in resisting the possibility of a supernatural entity than discovering the true nature of reality. Darwin writes in his biography: "Reason tells me of the extreme difficulty or rather impossibility of conceiving this immense and wonderful universe, including man with his capability of looking far backward and far into futurity, as the result of blind chance or necessity. When thus reflecting I feel compelled to look to a First Cause having an intelligent mind in some degree analogous to that of man; and I deserve to be called a Theist."[267]

In this postmodern age, where everything is relative, "truth"

---

264 Watson, "Adaptation," *Nature*, 233.
265 Todd, "A view from Kansas," Nature, 423.
266 Lewontin, "Billions and billions," *The New York Review*, 31.
267 Darwin, *The Autobiography*, 92–93.

often gets distorted, and many are tempted to look through the wrong pair of glasses. It would be tempting to try our different glasses, or worldviews, and choose the one which you find most comfortable. But, rather than settling for such a subjective approach, the three tests provide an alternative, and more objective, ways of assessing worldviews and working out which is the most credible.

ANOTHER COSMIC EVENT

How might an intelligent Creator demonstrate most clearly that our scientific understanding does not explain everything? The Bible suggests that He demonstrated this most powerfully through the resurrection of Jesus Christ. In Jesus Christ, divinity and humanity were mysteriously combined and, after He died, He rose again, giving hope to a dying world. His resurrection seems designed to be a demonstration of the supernatural.

Alongside the origin of the universe and the origin of life, the resurrection of Jesus Christ is the third event of cosmic significance and the fulcrum of the Christian faith. Paul writes that, without the resurrection of Christ, Christian faith is in vain (1 Cor. 15:14). That is why those who challenge Christian belief may naturally ridicule the very heart of the Christian gospel, the resurrection. When the British Magazine *The Spectator* asked Richard Dawkins whether he believed in the resurrection of Jesus Christ, he said: "People believe in the Resurrection, not because of good evidence (there isn't any) but because, if the Resurrection is not true, Christianity becomes null and void, and their life, they think, meaningless."

Writers who are unwilling to believe in the historic resurrection of Jesus Christ have proposed many theories to discount this Christian claim, including: the Swoon theory (Jesus fainted on the cross due to shock and loss of blood and later in the tomb somehow He got revived, managed to unwrap His dressings and roll the huge stone and climb out of the sealed tomb), the Hallucination theory (many people who saw the resurrected Jesus were just hallucinating), the Conspiracy theory (Christ's disciples stole the body of Jesus and fabricated the story) and so on. Though these theories attempted to portray the Resurrection of Jesus Christ as a fraud, the evidence strongly points towards the historicity of the Resurrection of Jesus Christ. Many books including the book by the investigative journalist an atheist, Lee Strobel who went out to disprove the new-found faith of his wife and finally ended up as a Christian provide valid evidence for the historical resurrection of Jesus Christ.[268] Strobel says that the gospel reports the resurrection in sober language, with specificity, and within a historical context that can be checked out. Also, a journalist called Frank Morison set out on a journey to disprove Christian claims about the resurrection of Jesus Christ, but midway, due to overwhelming evidence pointing to the resurrection; he became a Christian and wrote the book—*Who Moved the Stone?*

John 20 gives a detailed report of Jesus' resurrection. When Mary Magdalene came early in the morning, she found the stone rolled away from Jesus' tomb, and so she ran to Peter and another disciple (probably John) and said that somebody had taken away the Lord Jesus from the tomb. Peter and John ran together to the tomb and found something very profound – perhaps the turning-point in the then understanding of those two disciples. They saw (Greek word *theoreo*—to gaze upon, contemplate, consider)[269] that the napkin (Greek word *soudarion* — head-covering for the dead)[270] that had been placed over Jesus' head, was no longer with the linen cloths, but was now rolled up (Greek word *entulisso*—to roll up; not flattened)[271] in a separate place. They saw that the head covering of Jesus was not folded, but rolled up—presumably in the shape of a sphere—but with no head in it any longer. Even the other burial linen clothes were lying undisturbed with no body inside it. All this suggests that, after the resurrection, Jesus' body simply went through the clothes. The stone was probably moved by angels to let the disciples in and not to let Jesus out.

In John 20:15, when Jesus said to Mary Magdalene, "Woman, why are you weeping? Whom are you seeking?" she could not recognize Him and she perceived Him to be the gardener. Later that same day, Luke (Luke 24:13–35) gives an account of two disciples who met Jesus

---

268   Strobel, *The Case for Christ*, 2016.
269   *The Analytical Greek Lexicon of the NT*, s.v.
270   *The New Strong's Expanded Dictionary of Bible Words*, s.v.
271   *Thayer's Greek-English Lexicon of the NT*, s.v.

on the road to Emmaus, and verse 16 says that their eyes were kept from recognizing him. Again in John 21, we can read when Jesus showed Himself to His disciples at the sea of Tiberius, verse 4 says that the disciples did not know that it was Jesus while He stood on the shore. Why was the post-resurrection appearance of Jesus challenging to recognize for His close ones? We can infer that His body was not a mere physical body; it was a glorious body. He was not a mere spirit, because Jesus showed His disciples His hands and His sides and asked them to touch, emphasizing His bodily aspects.

Apostle John gave records of two distinct occasions (John 20:19–23, 26–29) when Christ in the body did appear before his disciples who were assembled behind closed doors for fear of the Jews. On both occasions, the account mentions that the doors were shut, yet Jesus came and stood in their midst. This does not mean that His body had been dematerialized, because Luke 24:39 clearly shows that Jesus' body was essentially material. Jesus says, "Look at my hands and my feet. It is I myself! Touch me and see; a ghost does not have flesh and bones, as you see I have." Jesus could eat physical food and also He had the crucifixion marks on His body (Luke 24:40–43; John 20:27). The text says that He came into the room when the doors were shut; it does not say that Jesus passed through the closed door. How could this have been possible? How did Jesus' body come out of the closed tomb and through the walls of a room? For a naturalistic scientist who is willing to accept the possibility that there are higher dimensions, it should be just as easy to believe that Jesus could get into a room with a closed door. How credible are multiple dimensions as compared with Jesus' resurrection body going through locked doors?

The novel Flatland of Edwin Abbott Abbott, describes the journeys of a Square, a mathematician and a resident of the two-dimensional Flatland.[272] They go to Spaceland (which has three dimensions), to Lineland (which has only one dimension) and even to Pointland (which has no dimensions). Ultimately, all this inspires ideas about visiting a land of four dimensions. Based on this novel, if a three-dimensional being living in Spaceland can lift a two-dimensional being living in flatland up out of the plane and then set him down

---

272 Abbott, *Flatland: A Romance*, 1884.

again in the plane after a day, his relatives would have given a missing person's report already. This action will appear miraculous in their eyes. In the same way, a three-dimensional being can enter the flatland and rob the houses of two-dimensional beings, without breaking or opening doors, while remaining invisible when he returns to his Spaceland. William Anthony Granville states, "A man (three-dimensional being) who has been translated from our space into a higher-dimensional space will remain invisible to earthly beings until he returns again to our space."[273] Any advanced higher-dimensional being will be unidentifiable and hidden to our senses until the higher-dimensional being chooses to reveal itself, in a way that the lower dimensional beings can perceive. If a higher dimensional being can remain invisible while intervening in lower dimensions, how much more can the eternal, glorious God, who is omniscient (all-knowing), omnipotent (all-powerful), omnibenevolent (all-good), and who is beyond all dimensions, intervene while choosing to remain invisible.

The resurrection of Jesus was not resuscitation—the reanimation of a dead body back to life, as before—but a cosmic event that transcends space and time. As the start of eternal life, it represents a complete victory over death. Intrinsic to this historic resurrection is the question of the 'after-life'; the ultimate question of destiny hangs on this tenet. The resurrection of Jesus Christ has tremendous transforming power; it was an earth-shaking experience for His disciples, and it had a profound effect on them, as they went everywhere boldly declaring the divinity of Jesus Christ, preaching His message to the world. The new, eternal life of the resurrection gives Jesus' followers the hope to face the present and also, through the promise of being with him in heaven, the hope to face the future.

"To see a World in a Grain of Sand, And a Heaven in a Wild Flower; Hold Infinity in the palm of your hand And Eternity in an hour."[274] We can see the big picture only from a minute piece in detail. The grain of sand is tiny, but the world is huge, but still a single grain reflects the massive world beautifully. Similarly, heaven is glorious and beautiful, whose reflection we can see in a little flower. It gives hope that, for a

---

273   Granville, *The Fourth Dimension*, 44.
274   Blake, *Auguries of Innocence*, eds. 1917.

believer of Jesus Christ, his eternity is ensured and he can be infinitely in the presence of his Creator, and when he sees the palm of his hand or hears the tick of a clock, they remind him every day of this incredible truth.

CONCLUSION

This research on the origin of the universe and the origin of life has revealed to me that the weight of evidence for an intelligent Designer is overwhelming and the mind boggles to grasp the minute details of this design. The beauty of the Designer is apparent as one looks inwards towards DNA or outwards towards our sun, our neighboring planets, our galaxy, and so on. I found that there is not necessarily any conflict between being a rigorous, analytical scientist and, at the same time, believing in a supernatural Being—a God who created the universe and yet also takes a personal interest in each one of us. I have presented evidence in favor of the truth claims of the biblical worldview based on cumulative results and have built my case as a lawyer files his arguments. The argument is not *a posteriori* argument, but a cumulative epistemological approach, based on the idea that the existence of God is more probable than the alternative explanations for the universe. I believe that the findings that I have presented here, based on reliable and compelling evidence, will help the readers to decide truth from falsehood with a reasonably clear and unbiased perspective. In the words of Blake, "If the doors of perception were cleansed everything would appear to man as it is."[275] In our age, there are many competing claims to truth, and many people even question whether there is such a thing as truth. Nevertheless, the reason that you are still holding this book in your hand is for sure, that you are interested in discovering the truth.

> the weight of evidence for an intelligent Designer is overwhelming

---

275 Blake, *The Marriage of Heaven and Hell*.

POSTSCRIPT

What made me write this piece of work? At one time in my life, helplessness and depression overpowered me leading to a negative perception about God and life. When I was walking towards the path of suicide, with my fist against God, God met me in a dramatic encounter, while I was dying of pneumonia and was on a ventilator. I had a vision of the Cross, with the holy blood from it covering my sins, and three times I heard a voice say, "I am giving you a new life." That vision changed my life. Though many questioned the validity of my vision, they could not deny the transformation, as they saw a completely transformed person in me, abounding with joy and peace. That is the power of meeting the ultimate reality—you will never be the same again.

The most important characteristic of a transformed life is to be at rest, and that is the importance of the seventh day of rest. Augustine said this about his own experience with restlessness: "Our heart is restless until it finds its rest in God." This is the greatest truth that I have discovered during this journey of my life. Entering into a complete rest in God is a beautiful process. It is a journey that takes an entire lifetime. It is not a one-time affair. God takes us through various steps to achieve this. A person who has entered into this rest walks with God every day of his or her life. What an incredible thing to possess!

After my dramatic encounter with Jesus Christ, I started a new life hand in hand with Him, drenched in His awesome presence that took over my whole being. I remember—once I was enjoying His presence while traveling in a vehicle that I did not even realize that my vehicle had got into an accident. When I opened my eyes, I saw that everybody had almost got down, and the windshield glass splattered everywhere. However, as the years passed by, I became quite busy with my own life, and this presence slowly ebbed, while the longing within me increased. I was restless and had the insatiable desire and longing to tear away the veil between God and me. I had that same feeling of lostness and pain when my second son got lost as a little boy. He was a toddler, and he followed another child and went missing, while I was busy shopping. After a frantic search, I finally found him near a shop. The deepest part of my heart longed to be bound together with God in some heroic purpose. I was unmoved by normal church spirituality, where everyone puts on that holy look as if they had resigned themselves to the life of a blank-faced monk in a remote wilderness. I used to be frustrated and felt like I was on a lifelong treasure hunt, every time some spiritual guy came along and told me that his life was a continual stream of miracles. Faith became a heavyweight and a great burden of guilt on me. Yet, I also thought of it as a mysterious thing that can change anything.

I often wondered whether a routine prayer, a drive to the office for work and then the mundane work at home, is all there is to life. I know that something was missing within my soul. There was a deep passionate voice within me that interrupted my mundane existence, creating a longing for adventure and intimacy. I tried to douse this voice with my work, but still, the voice was catching up with me so faintly whenever it caught me alone. I thought it was unspiritual; I ignored and tried to hide the deepest yearnings and restlessness of my heart. I kept myself busy with various assignments, separating and caging my heart. As a scientist by profession for the last three decades, I even tried to find a scientific answer to this wooing of the heart: Was it just a mechanical impulse or was my soul trying to send a message?

I did not understand that my story was interwoven in the adventure God was taking me on and that the restlessness and emptiness I felt did have something to do with God. Jesus was calling me through the longing of my heart and was trying to speak to me through its thirsts. He called me more deeply to the romance set within me and, thankfully I did not give up this romance but followed it up to its source—God. One calm morning, He spoke to me. I could hear His voice deep within, that He needed me fully. I understood that my heart had been distracted by many things, not knowing that this is the only One whom I need. Though the process was slow, and it took a long, difficult time to make Him alone the reason for my life and to give Him the highest priority in every way, it was worth the effort. He gave me complete rest, by becoming my intimate friend who speaks and listens to me.

Jesus brought me close to His heart through this journey of my restlessness and to discover that He alone is my soul's deepest desire through His warm, alive and poignantly haunting voice. I experienced a great breakthrough when I understood the mystery of true rest, by putting my complete trust and faith in God. As I changed the focus away from myself, my loneliness, my emptiness, and shifted the focus on to His greatness, His faithfulness, the pangs within my heart melted away, and I was able to enjoy this rest. I became like a weaned little child, quietly resting in the arms of God. There was no need to desperately cling, holding on to Him for dear life, but due to the developed trust and confidence, I learned that I can now rest, being rooted and established in His love. The insecurity in my relationship with God that had needed constant reassurance and affirmation, in the sheltered emotional stage, vanished. I am loved more than I can ever imagine. This art of resting made me constantly live in the presence of God, in the knowledge of God's continual love and to receive with an open heart what He has to do in my life.

The mathematical beauty and patterns inherent in nature also invoked a sense of awe and wonder within me and motivated me towards research to find the source of all the beauty and elegance of the universe. I found this search most rewarding. Writing this book was not a difficult task for me, as I could rest in Him and be enthralled at the various mysteries of His universe. I believe that, along with intellectual pursuit, each one of us needs to pursue our soul's desire to find the source of all longings and adventure. Scientists are on a continuous search for evidence for many mysterious phenomena: Where are the elusive Higgs bosons? What is the nature of dark matter and dark energy? Why is gravity so comparatively weak? I hope inquisitive people like them will also search for—and find—the Supernatural Being who is behind all these mysteries.

I sometimes think about the cross,
and shut my eyes, and try to see
the cruel nails and crown of thorns
and Jesus crucified for me.

But even could I see him die,
I could but see a little part
of that great love, which, like a fire,
is always burning in his heart.

It is most wonderful to know
his love for me so free and sure,
but 'tis more wonderful to see
my love for him so faint and poor.

And yet I want to love thee, Lord.
O light the flame within my heart,
and I will love thee more and more,
until I see thee as thou art.

William Walsham How

Bibliography

Albert, David. "On the Origin of Everything–A Universe From Nothing by Lawrence M. Krauss." *The New York Times*, March 23, 2012. https://www.nytimes.com/2012/03/25/books/review/a-universe-from-nothing-by-lawrence-m-krauss.html.

Alex, Bridget. "Why we're closer than ever to a timeline for human evolution." *The Guardian*, Dec. 22, 2016. https://www.theguardian.com/science/blog/2016/dec/22/why-were-closer-than-ever-to-a-timeline-for-human-evolution.

Alper, Joe. "Rethinking Neanderthals." *Smithsonian magazine*, Accessed March 5, 2012. http://www.smithsonianmag.com/science-nature/neanderthals.html.

Aristotle. *On the Heaven*. Translated by J. L. Stocks. Loeb Classical Library 338. Jan. 1970, first published 350.

Bagla, Pallava. "India's education minister assails evolutionary theory calls for curricula overhaul." *Science Magazine*, Jan. 22, 2018. http://www.sciencemag.org/news/2018/01/india-s-education-minister-assails-evolutionary-theory-calls-curricula-overhaul.

Bailey, Regina. "The 6 Kingdoms of Life." *ThoughtCo*, April 17, 2018. https://www.thoughtco.com/six-kingdoms-of-life-373414.

Barras, Colin. "Who are you? How the story of human origins is being rewritten." *New Scientist*. Aug. 23, 2017. https://www.newscientist.com/article/mg23531400-500-who-are-you-how-the-story-of-human-origins-is-being-rewritten/

Barnes, R.S.K. et al. *The Invertebrates: A New Synthesis*. Cambridge: Cambridge University Press, 1993.

Bavinck, Herman. *The Doctrine of God*. UK: The Banner of Truth Trust, 1996.

Behe, Michael J. *Darwin's Black Box: The Biochemical Challenge to Evolution*. Simon & Schuster Inc, 1996.

Bengtson, Stefan et al. "Three-dimensional preservation of cellular and subcellular structures suggests 1.6 billion-year-old crown-group red algae." *PLOS*, March 2017. http://journals.plos.org/plosbiology/article?id=10.1371/journal.pbio.2000735.

Bell, John S. "On the Einstein-Podolsky-Rosen Paradox." *Physics* 13 (1964) 195–200.

Bernhardt, Harold S. "The RNA world hypothesis: the worst theory of the early evolution of life." *BiolDirect*. July 13, 2012. https://www.ncbi.nlm.nih.gov/pmc/articles/PMC3495036.

Blake, William. *Auguries of Innocence*. Nicholson & Lee eds., 1917.

———.*The Marriage of Heaven and Hell*. Dover. 1994 first published 1790.
Boardman-Pretty, Freya. "First evidence that dinosaurs ate birds." *New Scientist*. Nov. 21, 2011.
Bohr, Niels. "Essays 1932-1957 on Atomic Physics and Human Knowledge." The Philosophical Writings of Neils Bohr, Vol.2.
Borwein, Jonathan, and David H. Bailey. "When science and philosophy collide in a 'fine-tuned' *Universe*." *Physics*. April 3, 2014. phys.org/pdf315737428.pdf.
Boyer, Steven D, and Christopher A. Hall. *The Mystery of God: Theology for Knowing the Unknowable*. Baker Academic, 2012.
Bradt, Steve. "3 Questions: Alan Guth on new insights into the 'Big Bang'." *MIT News*, March 20, 2014. http://news.mit.edu/2014/3-q-alan-guth-on-new-insights-into-the-big-bang.
Bruck, CarstenVanDe, and JurgenMifsud. "Searching for dark matter-dark energy interactions." *Phys. Review*. DOI: 10.1103/PhysRevD.97.023506.
Brush, Nigel. *The Limitations of Scientific Truth: Why Science Can't Answer Life's Ultimate Questions*. Kregel, 2005.
Budd, Graham E. "The earliest fossil record of the animals and its significance." Phil. Trans. of the Royal soc. B Biol Sci. 363 1496 (2008) 1425–1434. https://doi.org/10.1098/rstb.2007.2232.
Cain, Fraser. "What is Nothing?."*Universe Today*, Aug. 22, 2014. https://phys.org/news/2014-08-what-is-nothing.html
———. "When was the first light in the universe?." *Physics*, Nov. 7, 2016. https://phys.org/news/2016-11-universe.html
Castelvecchi, Davide. "Feuding physicists turn to philosophy for help." *Nature* 528 7583 (2015) 446–447.
———. "Is String Theory Science?." *Nature*. Dec. 23, 2015. https://www.scientificamerican.com/article/is-string-theory-science/
Chang, Kenneth. "In Explaining Life's Complexity Darwinists and Doubters Clash." *NY Times*. Aug. 23, 2005. https://archive.nytimes.com/www.nytimes.com/learning/teachers/featured_articles/20050823tuesday.html.
Chown, Marcus. "Forget dark matter–embrace my MOND theory instead." *New Scientist*, April 30, 2014. https://www.newscientist.com/article/mg22229670-400-forget-dark-matter-embrace-my-mond-theory-instead/
Cline, Austin. "Whale Pelvis: What Vestigial Organs Say About Evolution." *ThoughtCo*, Mar. 6, 2017. https://www.thoughtco.com/vestigial-organs-say-about-evolution-249897.
"CNN: Albert Einstein's colossal mistake." *Science News*, Nov. 11, 2015. http://www.egyptindependent.com/cnn-albert-einstein-s-colossal-mistake/
Collins, Robin. "The Fine-tuning Design Argument." In *Reason for the Hope Within*, edited by Michael Murray, Grand Rapids: Eerdmans, 1999.
Colson, Charles, and Nancy Pearcey. "How Now Shall We Live?." Tyndale House, 1999.
Conners, Deanna. "Why Earth has 4 seasons." *Earth Sky*, Sept. 20, 2016. http://earthsky.org/earth/can-you-explain-why-earth-has-four-seasons.
"A Cosmic Controversy Scientific American." *Scientific American*. Feb. 2017. https://blogs.scientificamerican.com/observations/a-cosmic-

controversy
Crick, Francis. *The Astonishing Hypothesis–The Scientific Search for the Soul.* London: Simon and Schuster, 1994.
Crockett, Christopher. "How did Earth get its water?." *Science News,* May 6, 2015. https://www.sciencenews.org/article/how-did-earth-get-its-water.
Dalton, Rex. "The fish that crawled out of the water." *Nature* April 5, 2006. https://www.nature.com/news/2006/060403/full/news060403-7.html.
Danchin, Antoine. "From chemical metabolism to life: the origin of the genetic coding process." *NCBI* June 2017. https://www.ncbi.nlm.nih.gov/pmc/articles/PMC5480338/
Darwin, Charles. *The Autobiography of Charles Darwin.* London: John Murray, 1887.
———.Francis, ed. *The life and letters of Charles Darwin.* Vol. 1 London: John Murray, 1887.
———.*The Origin of Species.* Chapter IX, 1859.
"Dark Energy Dark Matter." *NASA.* https://science.nasa.gov/astrophysics/focus-areas/what-is-dark-energy.
"Dark Matter Goes Missing in Oddball Galaxy." *NASA,* Mar. 28, 2018. https://www.nasa.gov/feature/goddard/2018/dark-matter-goes-missing-in-oddball-galaxy.
Das, Saswato R. "How the Higgs Boson Might Spell Doom for the Universe." *Scientific American,* March 26, 2013.
"Darwin's Greatest Challenge Tackled: The Mystery Of Eye Evolution." Science *Daily.* Nov 1, 2004. https://www.sciencedaily.com/releases/2004/10/041030215105.htm.
Dávalos, Liliana M. et al. "Understanding phylogenetic incongruence: lessons from phyllostomid bats." *NCBI,* Aug 14, 2012. https://www.ncbi.nlm.nih.gov/pmc/articles/PMC3573643.
Davies, Paul. "The Cosmos Might Be mostly devoid of life." *Scientific American.* Sept. 1, 2016. https://www.scientificamerican.com/article/the-cosmos-might-be-mostly-devoid-of-life/
———. "The origin of life. II: How did it begin?." *Science Progress* 84 1 (2001) 17.
———. "Physics and the mind of God: The Templeton Prize Address." *First Things,* Aug. 1995. https://www.firstthings.com/article/1995/08/003-physics-and-the-mind-of-god-the-templeton-prize-address-24.
———. "Yes the universe looks like a fix. But that doesn't mean that a god fixed it." *The Guardian,* June 26, 2007. https://www.theguardian.com/commentisfree/2007/jun/26/spaceexploration.comment.
———. *Cosmic Blueprint: New Discoveries In Natures Ability To Order Universe.* Simon & Schuster, 1988.
———.*The Fifth Miracle: The Search for the Origin and Meaning of Life.* Simon & Schuster, 1999.
Dawkins, Richard. *A Devil's Chaplain: Selected Essays.* Orion, 2004.
———.*The Blind Watchmaker: Why The Evidence Of Evolution Reveals A Universe Without Design.* W.W. Norton and Company, 1986.
De Bodt S, Maere S, and Van de Peer Y. "Genome Duplication And The Origin Of Angiosperms." *Trends In Ecology & Evolution* 20 11 (2005) 591–597.
Degnan, James H, and Noah A Rosenberg. "Gene tree discordance phylogenetic

inference and the multispecies coalescent." *Trends in Ecology and Evolution* 24 6 (2009) 332–340.

Demuth, Jeffery P. et al. "The Evolution of Mammalian Gene Families." *PLOS*, Dec. 20, 2006. http://journals.plos.org/plosone/article?id=10.1371/journal.pone.0000085.

Dickerson, Kelly. "Stephen Hawking Says 'God Particle' Could Wipe Out the Universe." *Live Science*, Sept. 8, 2014.

———. "This bizarre experiment just produced the best evidence yet of the universe's 'spooky' side." *Business Insider*, Oct. 28, 2015. https://www.businessinsider.in/This-bizarre-experiment-just-produced-the-best-evidence-yet-of-the-universes-spooky-side/articleshow/49568014.cms.

Donoghue, Philip C. and, Ziheng Yang, "The evolution of methods for establishing evolutionary timescales," *Phil. Trans. of the Royal Society B* (July 2016). https://doi.org/10.1098/rstb.2016.0020. https://doi.org/10.1098/rstb.2016.0020.0020.

Doolittle, W Ford. "The practice of classification and the theory of evolution and what the demise of Charles Darwin's tree of life hypothesis means for both of them." *Phil. Trans. of the Royal Society B: Biological Sciences* 364 (2009) 2226. doi:10.1098/rstb.2009.0032.

Dorminey, Bruce. "Without the Moon Would There Be Life on Earth?." *Scientific American.* April 21, 2009. https://www.scientificamerican.com/article/moon-life-tides/

Douglas, Kate. "Asia's mysterious role in the early origins of humanity." *New Scientist.* July 4, 2018. https://www.newscientist.com/article/mg23931850-200-asias-mysterious-role-in-the-early-origins-of-humanity/

Dunn, James D.G. *The Cambridge Companion to St Paul.* Cambridge Univ. Press, 2003.

Dunning, Hayley. "The Big Bang might have been just a Big Bounce." *Physics.* July 12, 2016. https://phys.org/news/2016-07-big.html.

Dyson, Freeman. *Disturbing the Universe.* New York: Harper & Row, 1979.

Eastman, Mark, and Chuck Missler. "The Origin of Life and The Suppression of Truth." http://xwalk.ca/origin.html.

"Einstein and the EPR Paradox." *APS News.* 14 10 (2005). https://www.aps.org/publications/apsnews/200511/history.cfm.

Elias, Economou. "An Orthodox View of the Ecological Crisis." *Theologia* 61 4 (1990) 607–619.

Elliott, Sober. *Ockam's Razor: A User's Manual.* Cambridge University Press, 2015.

Erich, Becher. "The Philosophical Views of Ernst Mach." *The Philosophical Review.* 14 5 (1905) 535–562. https://www.jstor.org/stable/pdf/2177489.pdf

Ewert, Winston. "The dependency graph of life." *BIO-Complexity.* 3 (2018):1–27. doi:10.5048/BIO-C.2018.3.

Fazekas, Andrew. "Life Ingredients Found in Superhot Meteorites." *National Geography News,* Dec. 21, 2010. https://news.nationalgeographic.com/news/2010/12/101220-asteroid-meteorite-life-space-science/

Feduccia, Alan. *The Origin and Evolution of Birds.* 2nd ed. Yale University Press, 1999.

———. "Birds are dinosaurs: simple answer to a complex problem." *Auk* 119 No.

4 (2002) 1187–1201.
Flatland, Edwin A. Abbott. "A Romance of Many Dimensions." Dover Thrift Editions, 1884.
Folger, Tim. "Einstein's Grand Quest for a Unified Theory."*Discover Magazine*, Sept. 30, 2004.
Gauger, Ann, Douglas Axe, and Casey Luskin. *Science and Human Origins.* Discovery Institute, 2012.
Gee, Henry. "Return to the planet of the apes." *Nature* 412 (2001) 131–32.
Gefter, Amanda. "Concept of 'hypercosmic God' wins Templeton Prize." *New Scientist*. March 16, 2009.
Getonthetrain. "Boskop Man: Ancient Humans With Extraordinarily Large Skulls—How Smart Were They?." *Steemit*. https://steemit.com/history/@getonthetrain/boskop-man-ancient-humans-with-extraordinarily-large-skulls-how-smart-were-they.
Gesteland, Ray and John Atkins, "The RNA World." Monograph, vol. 24, 1993.
Ghose, Tia. "Did a Volcano Wipe Out the Neanderthals?" *Live Science*. Dec. 30, 2014. https://www.livescience.com/49290-volcano-did-not-kill-neanderthals.html.
———. "Mesozoic Era: Age of the Dinosaurs." *Live Science*. Jan. 7, 2015. https://www.livescience.com/38596-mesozoic-era.html.
———. "Origin of Life: Did a Simple Pump Drive Process?." *Live Science*. Jan. 11, 2013. https://www.yahoo.com/news/origin-life-did-simple-pump-drive-process-154410143.html.
Gibbons, Ann. "Bonobos Join Chimps as Closest Human Relatives." *Science Magazine*. June 13, 2012.
———. "Which of Our Genes Make Us Human?." *Science* 281 5382 (1998)1432–1434.
Gibney, Elizabeth. "Dark-Matter Hunt Fails to Find the Elusive Particles." *Nature* 551 (2017). https://www.nature.com/polopoly_fs/1.22970!/menu/main/topColumns/topLeftColumn/pdf/551153a.pdf.
Gödel, Kurt. "Some mathematical results on completeness and consistency On Formally Undecidable Propositions of Principia Mathematica and related systems iand on completeness and consistency." In Two Fundamental Texts in Mathematical Logic edited by J. van Heijenoort (ed). (1930b, 1931 and 1931a).
Dewitt, Bryce Seligman, and Neill Graham.(eds.). *The Many-Worlds Interpretation of Quantum Mechanics*. Princeton NJ: Princeton University Press, 2015.
Granville, William Anthony. *The Fourth Dimension and the Bible*. Bridge-Logos. 1922.
Green, William Henry. *Primeval Chronology Bibliotheca*. Sacra 47 (1890) 286–303.
Greene, B. *The Elegant Universe: Superstrings Hidden Dimensions and the Quest for the Ultimate Theory*. New York: W.W. Norton & Company, 2010.
Greenleaf, Simon. *An Examination of the Testimony of the Four Evangelists By the Rules of Evidence Administered in Courts of Justice*. Boston: C. C. Little and J. Brown, 1846.
Hawking, Stephen. *A Brief History of Time: From Big Bang to Black Holes*. New York: Bantam Books, 1998.
———. "The Beginning of Time Lecture."1996. http://www.hawking.org.uk/

the-beginning-of-time.html.
———— and Leonard Mlodinow. *The Grand Design*. Bantam Books, 2011.
———— and Thomas Hertog. "Populating the Landscape: A Top Down Approach." Last modified February 10 2006 accessed December 24 2016. https://arxiv.org/PS_cache/hep-th/pdf/0602/0602091v2.pdf.
Heckman, Daniel et al. "Molecular evidence for the early colonization of land by fungi and plants." *Science* 293 5532 (2001) 1129–1133. http://science.sciencemag.org/content/293/5532/1129.
Heilbron, John Lewis. *The Sun in the Church: Cathedrals as Solar Observatories.* Harvard University Press, 2001.
Heisenberg, Werner. *The Physical Principles of the Quantum Theory*. Dover, 1980.
Hensenet, B. al. "Loophole-free Bell inequality violation using electron spins separated by 1.3 kilometres." Nature 526 (2015) 682- 686.
Ho, Simon. "The Molecular Clock and Estimating Species Divergence." *Nature Education* 1 (2008). https://www.nature.com/scitable/topicpage/the-molecular-clock-and-estimating-species-divergence-41971.
Hoffman, Hillel J. "The Permian Extinction—When Life Nearly Came to an End." *National Geographic Magazine.* https://www.nationalgeographic.com/science/prehistoric-world/permian-extinction/
Horgan, John. "Physicist Slams Cosmic Theory He Helped Conceive." *Scientific American.* Dec.1, 2014. https://blogs.scientificamerican.com/cross-check/physicist-slams-cosmic-theory-he-helped-conceive/.
Hoyle, Fred and Chandra Wickramasinghe. *Evolution from Space*. London: J.M. Dent and Sons, 1981.
"Hoyle on Evolution."*Nature* 294 5837 (1981) 105. https://www.nature.com/articles/294104b0.pdf.
"How Far Was The Moon From Earth When it Started Forming?."*Nevada's Space Site.*
Huffingtonpost. "Can We Measure the Earth's Age According to the Bible?." Updated Feb 15, 2018. https://www.huffingtonpost.com/quora/can-we-measure-the-earths_b_14748994.html.
Hume, David, L. A. Selby-Bigge ed. *A Treatise of Human Nature*. Dover, 2003.
Hunter, Cornelius G. "Darwin's Predictions." 2015. https://sites.google.com/site/darwinspredictions/
Ian, Sample. "Evolution: Charles Darwin was wrong about the tree of life." *The Guardian,* Jan 21, 2009. https://www.theguardian.com/science/2009/jan/21/charles-darwin-evolution-species-tree-life.
Ijjas, Anna, Paul Steinhardt, and Abraham Loeb. "Pop goes the Universe." *Scientific American.* Feb. 2017. https://physics.princeton.edu/~cosmo/sciam/assets/pdfs/SciAm.pdf
"An integrated encyclopedia of DNA elements in the human genome." *Nature* 489 9 (2012) 57–74.
Janvier, Philippe, and Gaël Clément. "Muddy tetrapod origins." Nature 4631 (2010) 40–41.
Jastrow, Robert. *Until the Sun Dies*. W.W. Norton, 1977.
Jeyachandran, L.T. "Science: God's friend or foe." *Jubilee.* https://d1nwfrzxhi18dp.cloudfront.net/uploads/resource_library/attachment/file/56/20100401_Jubilee01_ScienceandChristianity_JoeBoot-Editor.pdf.

Joshi, Sonali S. "Origin of Life: The Panspermia Theory Helix." Dec 2, 2008. https://helix.northwestern.edu/article/origin-life-panspermia-theory.

Kakuo, Shingo, K. Asaoka, and T. Ide. "Human Is a Unique Species Among Primates in Terms of Telomere Length." *Biochem. Biophys. Res. Commun.* 263 (1999) 308–314.

Kaluza, Th. "On the Unification Problem in Physics." Int. J. Mod. Phys. D, Vol. 27, No. 14 (2018) 187001 (trans) In Math. Phys.1921, 966-972. DOI: 10.1142/S0218271818700017

Kaufman, Marc. "Did Life on Earth Come From Mars?."*National Geographic News.* Sept. 5, 2013.

Kayebonto. "Scientists Abandon the Oscillating Universe Theory." https://www.scribd.com/document/90139627/Scientists-Abandon-the-Oscillating-Universe-Theory

Kiefer, C. *Quantum Gravity.* 3rd ed. Oxford UK: Oxford University Press, 2012.

Klein, O. "The atomicity of electricity as a quantum theory law." *Nature* 118 (1926) 516.

Koberlein, Brian. "Second Light in Cosmology." *Cosmology.* Feb. 7, 2015. https://briankoberlein.com/2015/02/07/second-light.

Larson, Richard B. "The First Stars in the Universe." *Scientific American.* Jan. 19, 2009. https://www.scientificamerican.com/article/the-first-stars-in-the-un/

Lawton, Graham. "Why Darwin was wrong about the tree of life." *New Scientist.* Jan. 21, 2009. https://www.newscientist.com/article/mg20126921-600-why-darwin-was-wrong-about-the-tree-of-life/

Lemonick, Michael D. "A Bit of Neanderthal in Us All?." *Time Magazine*, April 25, 1999. http://content.time.com/time/magazine/article/0,9171,23543,00.html.

———. "Why Einstein was wrong about being wrong." *Los Angeles Times.* Oct. 14, 2011. http://articles.latimes.com/2011/oct/11/opinion/la-oe-1011-lemonick-einstein-20111011.

——— and Andrea Dorfman. "Ardi Is a New Piece for the Evolution Puzzle." *TIME.* Oct. 1, 2009. http://content.time.com/time/magazine/article/0,9171,1927289,00.html.

Lennox, John. "Intellectually Fulfilling Faith: Lessons from C.S. Lewis." *Issue of Broadcast Talks*1 3 (2016).http://www.cslewisinstitute.org/Intellectually_Fulfilling_Faith_Lessons_from_C_S_Lewis

———. *Seven Days That Divide the World: The Beginning According To Genesis & Science.* Zondervan, 2011.

Lewin, Roger. *Bones of Contention: Controversies in the Search for Human Origins.* New York: Simon and Schuster, 1987.

Lewis, C.S. *Mere Christianity.* Harper Collins, 1952.

Lewontin, R. "Billions and billions of demons." *The New York Review.* Jan. 9, 1997.

Leydesdorff, L. "Can scientific journals be classified in terms of aggregated journal-journal citation relations using the Journal Citation Reports?." *Journal of the American Society for Information Science and Technology* 57 (2006) 601–613.

"Our Living Planet Shapes the Search for Life Beyond Earth." *NASA*, Nov. 16, 2017. https://www.nasa.gov/feature/jpl/our-living-planet-shapes-the-

search-for-life-beyond-earth.
Luskin, Casey. "Problem 10: Neo-Darwinism's Long History of Inaccurate Predictions about Junk Organs and Junk DNA." *Evolution news*. Feb. 19, 2015. https://evolutionnews.org/2015/02/problem_10_neo/.
Lynch, Gary and Richard Granger. "What Happened to the Hominids Who May Have Been Smarter Than Us?." *Discover magazine*, Dec. 28, 2009. http://discovermagazine.com/2009/the-brain-2/28-what-happened-to-hominids-who-were-smarter-than-us.
McRae, Michael. "Neil Armstrong." *ASME* July 2012. https://www.asme.org/engineering-topics/articles/aerospace-defense/neil-armstrong
Malassé and, Anne Dambricourt. "The first Indo-French Prehistorical Mission in Siwaliks and the discovery of anthropic activities at 2.6 million years." *Science Direct* 15 3–4 (2016) 281–294.
Mangalwadi, Vishal. *Truth and Transformation: A Manifesto for Ailing Nations.* YVAM, 2009.
Margenau, Henry, ed. and Roy Abraham Varghese. *Cosmos Bios, Theos: Scientists Reflect On Science God And The Origins Of The Universe Life And Homo Sapiens*. La Salle Ill: Open Court, 1992
Marsolek, Patrick. "The Goldilocks Hypothesis." 2016. https://atlantisrisingmagazine.com/article/the-goldilocks-hypothesis/
Max, Planck and James Vincent Murphy, ed. *Where is Science Going?* New York: AMS, 1932.
———. *Signature in the Cell: DNA and the evidence for Intelligent Design.* Harper One, 2009.
———. *Darwin's Doubt: The Explosive Origin of Animal Life and the Case for Intelligent Design*.rev. ed. Harper One, 2014.
Miller, Brian. "BIO-Complexity Presents Better Model than Common Ancestry for Explaining Pattern of Nature." *Evolution news*. July 19, 2018. https://evolutionnews.org/2018/07/bio-complexity-presents-a-better-model-than-common-ancestry-for-explaining-the-pattern-of-nature
———. "*Eye Evolution*: A Closer Look." *Evolution News*, Feb. 13, 2017. https://evolutionnews.org/2017/02/eye_evolution_a/
———. "Eye Evolution: The Waiting Is the Hardest Part." *Evolution News*. Feb. 15, 2017. https://evolutionnews.org/2017/02/eye_evolution_t/
Larralde, R., Robertson MP, and Miller SL. "Rates of decomposition of ribose and other sugars: Implications for chemical evolution." *PNAS* 92 (1995) 8158–8160. https://www.ncbi.nlm.nih.gov/pubmed/7667262.
Mattila, Tiina M, and Folmer Bokma. "Extant mammal body masses suggest punctuated equilibrium." *Proc. R. Soc. B: Biol. Sci.* 275 (2008) 2195.
Margenau, Henry, and Roy A. Varghese, eds. *Cosmos, Bios, Theos: Scientists Reflect on Science, God, and the Origins of the Universe, Life, and Homo Sapiens.* La Salle Ill.: Open Court Pub. Co., 1992.
Mitchell, Jacqueline. "In the Beginning Was the Beginning Cosmologist Alex Vilenkin does the math to show that the universe indeed had a starting point." *Tufts Now*. May 29, 2012. https://now.tufts.edu/articles/beginning-was-beginning.
Moreland, J.P et al., *Theistic Evolution: A Scientific, Philosophical, and Theological Critique*. Wheaton, Illinois: Crossway, 2017.

Mosher, Dave. "Life on Earth Began on Land Not in Sea?." *National Geographic News*. Feb. 15, 2012. https://news.nationalgeographic.com/news/2012/02/120213-first-life-land-mud-darwin-evolution-animals-science/

Moskowitz, Clara. "Gravity Waves from Big Bang Detected." *Scientific American*, March 17, 2014. https://www.scientificamerican.com/article/gravity-waves-cmb-b-mode-polarization/

Nelson, Paul. "Beautiful Monster —Theistic Evolution: A Scientific Philosophical and Theological Critique Is Here!" *Evolution news*, Nov. 2, 2017.

———, and Jonathan Wells, "*Homology in Biology*: Problem for Naturalistic Science and Prospect for Intelligent Design," In JA Campbell, Stephen C Meyer, eds., *Darwinism, Design, and public Education*. 2003.

Nickels, Martin. "The Nature of Modern Science & Scientific Knowledge." *Illinois State University*. Aug. 1998. http://www.indiana.edu/~ensiweb/mart.nos.pdf

Niles, Eldridge. *The Myths of Human Evolution*. Columbia Univ Press, 1984.

"NIST Clock Experiment Demonstrates That Your Head is Older Than Your Feet." *NIST*. Sept. 28, 2010. https://www.nist.gov/news-events/news/2010/09/nist-clock-experiment-demonstrates-your-head-older-your-feet.

"No Universe without Big Bang." *Physics*, June 15, 2017. https://phys.org/news/2017-06-universe-big.html.

O'Neill, Iain. "Homozygosity for constitutional chromosomal rearrangements: a systematic review with reference to origin ascertainment and phenotype." *Journal of Human Genetics* 55 (2010) 559–564.

Orzel, Chad. *How to teach Quantum Physics to your Dog*. Scribner, 2010.

Overbye, Dennis. "LIGO Detects Fierce Collision of Neutron Stars for the First Time." *New York Times*. Oct. 16, 2017. https://www.nytimes.com/2017/10/16/science/ligo-neutron-stars-collision.html.

Owen, Bruce. "Darwin's big problem and Mendelian genetics." 2011. http://bruceowen.com/introbiological/a201-11f-07-DarwinsProblemMendel.pdf.

Owen, James. "Orangutans May Be Closest Human Relatives Not Chimps." *National Geographic News*. June 23, 2009. https://news.nationalgeographic.com/news/2009/06/090623-humans-chimps-related.html.

Pappas, Stephanie. "What is Gondwana?." *Live Science*, June 7, 2013. https://www.livescience.com/37285-gondwana.html.

Perkins, Sid. "Earth is only just within the Sun's habitable zone." *Nature*, Dec. 11, 2013. https://www.nature.com/news/earth-is-only-just-within-the-sun-s-habitable-zone-1.14353.

Phillips, Tony. "Cosmic Rays Hit Space Age High." *NASA*. https://www.nasa.gov/topics/solarsystem/features/ray_surge.html.

"Physics World Evidence for recent star formation seen at Milky Way's centre." *Physics World*, Mar. 17, 2015. https://physicsworld.com/a/evidence-for-recent-star-formation-seen-at-milky-ways-centre/

Pickrell, John. "Humans Chimps Not as Closely Related as Thought?." *National Geographic News*. Sept. 24, 2002. https://news.nationalgeographic.com/

news/2002/09/0924_020924_dnachimp.html
Plato. *The Allegory of the Cave*. The Republic Book VII.
Polanyi, Michael. "Scientific Beliefs." *Ethics* 61 1 (1950).
———.*The Tacit Dimension*. New York: Anchor Books, 1967.
Polkinghorne, John C. *Science and Creation: The Search for Understanding*. London: SPCK, 1988.
Popper, Karl. *The Logic of Scientific Discovery*. New York: Basic Books. 1959.
Premack, David. "Human and animal cognition: Continuity and discontinuity." *Proc Natl Acad Sci* 104 35 (2007) 13861–13867.
Prüfer, Kay et al. "The bonobo genome compared with the chimpanzee and human genomes." *Nature* 486 (2012) 527–531. https://www.nature.com/articles/nature11128.
Pulquério, Mário J., and Richard A. Nichols. "Dates from the molecular clock: how wrong can we be?." *Trends in Ecology& Evolution* 22 4 (2007)180–184.
Redd, Nola Taylor. "How Did the Solar System Form?" *Space* Jan. 31, 2017. https://www.space.com/35526-solar-system-formation.html
———. "How Was Earth Formed?." *Space* Oct. 31, 2016. https://www.space.com/19175-how-was-earth-formed.html.
———."Is dark energy caused by 'chameleon' particles?." *Space*. Aug. 23, 2015. https://www.csmonitor.com/Science/2015/0823/ Is-dark-energy-caused-by-chameleon-particles.
Ricardo, Alonso and Jack W. Szostak. "Origin of Life on Earth." *Scientific American* 301 (2009) 54–61. https://www.mcb.ucdavis.edu/faculty-labs/scholey/journal%20papers/ricardo-szostak-sa2009.pdf
"Richard Dawkins: You Ask the Questions Special." *Independent*, Dec. 4, 2006. https://www.independent.co.uk/news/people/profiles/richard-dawkins-you-ask-the-questions-special-427003.html
Rincon, Paul. "Neanderthals gave us disease genes." *BBC News*, Jan. 29, 2014. https://www.bbc.com/news/science-environment-25944817.
Robinson, Richard. "Jump-Starting a Cellular World: Investigating the Origin of Life from Soup to Networks." *NCBI*. Nov. 15, 2005. https://www.ncbi.nlm.nih.gov/pmc/articles/PMC1283399.
Ross, Hugh. *The Creator And The Cosmos*. Colorado Springs: Navpress, 2001.
Rufu, Raluca, Oded, and Hagai. "A Multiple-Impact Origin for the Moon." *Nature Geoscience* 10 (2017) 89–94.
Ruse, Michael and Joseph Travis. *Evolution: The First Four Billion Years*. Harvard University Press, 2009.
Sagan, Carl. *Pale Blue Dot: A Vision of the Human Future in Space*. Random House, 1994.
Sailhamer, John. *Genesis Unbound: A Provocative New Look at the Creation Account*. Multnomah Books, 1996.
Saladino, R. et al.."Genetics first or metabolism first? The formamideclue." *ChemSoc Rev*. June 8, 2012. https://www.ncbi.nlm.nih.gov/pubmed/22684046.
Schaeffer, Francis A. *How Should We Then Live? The Rise and Decline of Western Thought and Culture*. Old Tappan NJ: Fleming H. Revell, 1976.
Schrodinger, Erwin. "Discussion of probability relations between separated

systems." *Cambridge Philosophical Society Proceedings* 31 4 (1935) 555-563.

"Science Cannot Fully Describe Reality Says Templeton Prize Winner." *Science Magazine*. Mar. 16, 2009. http://www.sciencemag.org/news/2009/03/science-cannot-fully-describe-reality-says-templeton-prize-winner.

"Scientists discover hidden galaxies behind the milky way." *ICRAR*, Feb. 10, 2016. https://www.icrar.org/hidden-galaxies/

Shiwnarain, Mohendra. "Ancient Skull Found In China Could Change The History Of Humans." *Science Trends*. Nov. 24, 2017.https://sciencetrends.com/ancient-skull-found-china-change-history-humans/

Sire, James W. *The Universe Next Door: A Basic Worldview Catalog*. Downers Grove: InterVarsity Press, 2009.

Sokol, Joshua. "Volcanoes plus asteroid might have finished off dinosaurs." *New Scientist*. Oct. 1, 2015. https://www.newscientist.com/article/dn28275-volcanoes-plus-asteroid-might-have-finished-off-dinosaurs/

Stark, R. *For the Glory of God: How Monotheism Led To Reformations Science Witch-Hunts And The End Of Slavery*. Princeton University Press, 2003.

State, Penn. "Oxygen May Be Cause Of First Snowball." *Earth Science Daily*. Oct. 29, 1999. https://www.sciencedaily.com/releases/1999/10/991029071656.htm.

Sternberg, Richard. "On the Roles of Repetitive DNA Elements in the Context of a Unified Genomic– Epigenetic System." *Annals of the New York Academy of Sciences* 981 (2002) 154–188.

Strobel, Lee. *The Case For A Creator*. Grand Rapids, Mich.: Zondervan, 2004.

———. *The Case for Christ*. Zondervan, 2016.

Swinburne, Richard. *Is There a God?*. Oxford University Press, 1996.

Syvanen, Michael. "Evolutionary Implications of Horizontal Gene Transfer." *Annual Review of Genetics* 46 (2012) 341–358.

Tarlach, Gemma. "Everything Worth Knowing About . . . Scientific Dating Methods." *Discover Magazine*. June 01, 2016. http://discovermagazine.com/2016/jul-aug/scientific-dating-methods.

Tegmark, Max. "Is the Universe Made of Math?." *Scientific American*, Jan. 10, 2014. https://www.scientificamerican.com/article/is-the-universe-made-of-math-excerpt/.

Than, Ker. "New Type of Human Discovered via Single Pinky Finger Siberian fossil points to unknown migration out of Africa." *National Geographic News*. Mar. 25, 2010. https://news.nationalgeographic.com/news/2010/03/100325-new-human-species-x-woman-pinky-finger-denisova-dna-nature/.

———. "Fossils Could Force Rethink of Human Evolution. *Live Science*. Aug. 8, 2007. https://www.livescience.com/1756-fossils-force-rethink-human-evolution.html.

Thewissen, J. G. M. Et al. "Skeletons Of Terrestrial Cetaceans And The Relationship Of Whales To Artiodactyls." *Nature* 413 6853 (2001) 277-281.

Thompson, Helen. "Promiscuous Whales Make Good Use of Their Pelvises." *Smithsonian*. Sept. 8, 2014. https://www.smithsonianmag.com/science-nature/promiscuous-whales-make-good-use-pelvises-180952620.

Theobald, Douglas L. "A formal test of the theory of universal common

ancestry." *Nature* 465 (2010) 226.
The Times Eureka. "Excerpt from Stephen Hawking and Leonard Mlodinow's The Grand Design." Sept. 25, 2010.
Todd, S.C. "A view from Kansas on that evolution debate." *Nature,* 401 6752 (1999) 423. https://www.ncbi.nlm.nih.gov/pubmed/10519534.
Tomkins, Jeffrey, and Jerry Bergman. "The chromosome 2 fusion model of human evolution—part 2: Re-analysis of the genomic data." *Journal of Creation* 25 2 (2011)111–117.
Tryon, Edward P. "Is the Universe a Vacuum Fluctuation?." *Nature,* 246 (1973) 396–397. https://www.nature.com/articles/246396a0.
Vasas, Vera, Szathmáry.E and Santos M. "Lack of evolvability in self-sustaining autocatalytic networks constraints metabolism-first scenarios for the origin of life," *PNAS,* Jan. 2010. http://www.pnas.org/content/107/4/1470.
Vilenkin, Alex. *Many Worlds in One: The Search for Other Universes.* Hill and Wang, 2006.
Wall, Mike. "The Big Bang: What Really Happened at Our Universe's Birth?." *Space,* Oct. 21, 2011. https://www.space.com/13347-big-bang-origins-universe-birth.html.
Walling, Jeffrey L. *His Story: As a Matter of Fact.* iUniverse, 2017.
Walton, John H. *The Lost World of Genesis One: Ancient Cosmology and the Origins Debate.* IVP Academic, 2009.
Ward, Keith. *Why there almost certainly is a God: Doubting Dawkins.* London: Lion Hudson, 2008.
Watson, D.M.S. "Adaptation." *Nature* 123 (1929) 233.
Weinberg, S. *The First Three Minutes: A modern view of the origin of the Universe.* Basic Books. 1993. http://www.slobodni-univerzitet-srbije.org/files/weinberg-steven-the-first-three-minutes.pdf.
"What have we got in common with a gorilla? Insight into human evolution from gorilla genome sequence." *Science News.* March 7, 2012.
  https://www.sciencedaily.com/releases/2012/03/120307132210.htm.
Wheeler, John Archibald. *A Journey into Gravity and Space time.* Scientific American Library, Series 31, 1990.
Wheeler, John Archibald with Kenneth Ford. *Geons, Black Holes and Quantum Foam.* W. W. Norton & Co., 2000.
Whitcomb, John C., and Henry M. Morris. *The Genesis Flood.* Grand Rapids. MI: Baker, 1977.
Wilford, John Noble. "How the Whale Lost Its Legs and returned to the Sea." *The New York Times,* May 3, 1994. https://www.nytimes.com/1994/05/03/science/how-the-whale-lost-its-legs-and-returned-to-the-sea.html.
———. "The Human Family Tree Has Become a Bush With Many Branches." *The New York Times,* June 26, 2007. https://www.nytimes.com/2007/06/26/science/26ance.html.
Wiseman, P.J. *Creation Revealed in Six Days: The evidence of Scripture confirmed by Archaeology.* Marshall. Morgan & Scott Ltd, 1958.
Witten, Edward. "Fivebranes and M-theory on an orbifold." *Nucl. Phys.* 463 (1996) 383.
Wolchover, Natalie. "Have We Been Interpreting Quantum Mechanics Wrong This Whole Time?." *Quanta Magazine Science.* June 30, 2014 .https://

www.wired.com/2014/06/the-new-quantum-reality/
———. "What If There Were No Seasons?." *Live Science*. March 9, 2012. https://www.livescience.com/18972-earth-seasons-tilt.html.
Wood, Bernard, and Collard. "The human genus."*NCBI* 284 5411 (1999) 65–71. https://www.ncbi.nlm.nih.gov/pubmed/10102822.
Zacharias, Ravi. "Think again–Deep Questions Slice of Infinity." August 28, 2014. http://rzim.org/just-thinking/think-again-deep-questions/
Zhou, Fucheng and Zhang Zhonghe. "A Primitive Enantiornithine Bird and the Origin of Feathers." *Science* 290 (2000)1955–1959.
Zimmer, Carl. "The Surprising Origins of Evolutionary Complexity." *Scientific American*. Aug. 1, 2013. https://www.scientificamerican.com/article/the-surprising-origins-of-evolutionary-complexity/
Zorich, Zach. "Is "Junk DNA" What Makes Humans Unique?." *Scientific American*. Jan. 30, 2018. https://www.scientificamerican.com/article/is-junk-dna-what-makes-humans-unique/
Zukav, Gary. *The Dancing Wu Li Masters: An overview of the New Physics*. William Morrow and Co., 1979.

www.ingramcontent.com/pod-product-compliance
Lightning Source LLC
Chambersburg PA
CBHW070740160426
43192CB00009B/1514